The Dance of Growth: Temperature, Diffusion, and Seeding in Epitaxial Film Formation

Mackil

TABLE OF CONTENTS

Chapter 1: Introduction

1.1 2D Material Family

Recent research has demonstrated that in addition to the composition and arrangement of atoms in materials, dimensionality has profoundly impacted on determining their fundamental structure property-processing-performance relationships. This has been most strikingly manifested in the 2D layered materials which display exotic condensed-matter phenomena that are present only in a monolayer form.[1] From a semiconductor perspective, a decrease in film thickness to the ultimate limit of the atomic, sub-nanometer length scale, the holy grail for conventional semiconductors (such as Si), would be forthcoming and shed lights on ultrathin and flexible electronics, photovoltaics, sensor, and display technology. In this light, transition metals dichalcogenides (TMDs) with the chemical formula MX_2 (X= S, Se, Te, and M= transition metal), which can form stable, three-atom-thick monolayers, tantalizingly hold the prospect on providing semiconducting materials with high electrical carrier mobility. In parallel, their unique electronic band structures open up new inroads in further enhancing the functionalities of such devices through strain-induced bandgap modulation, substantial excitonic effect, layered dependent indirect-to-direct bandgap transition, in-plane piezoelectricity, and emerging valley electronics.[2, 3]

The most explored member of the TMDs family is MoS_2. The crystal structure is schematically represented in Figure 1.1. The Mo-S bonds are predominately covalent, whereas the sandwich layers are held by the weak van der Waals (vdW) interaction. The structural anisotropy, therefore, enables an exfoliation of monolayers where the physical and chemical characteristics are drastically different from their bulk counterparts. Aside from the layer-dependent evolution in electronic properties, TMDs exhibits different polymorphs, such as 1T, 2H, and 3R, as a result of different stacking of the M and X layers as illustrated in Figure 1.1c. Therefore, 2D TMDs can be utilized for a wide range of applications.

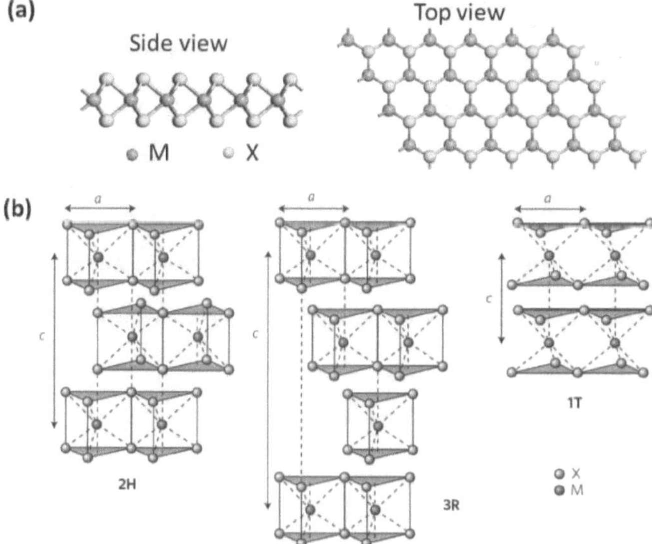

Figure 1.1 Schematics of the MX₂ structure. (a) Side view of the MX$_2$ model (left), and top view (right). (b) 2H (hexagonal symmetry, two layers per repeat unit, trigonal prismatic coordination), 3R (rhombohedral symmetry, three layers per repeat unit, trigonal prismatic coordination), and 1T (tetragonal symmetry, one layer per repeat unit, octahedral coordination). The chalcogenide atoms (X) are yellow and the metal atoms (M) are grey. The lattice constants a are in the range of 3.1 to 3.7 Å for different materials.[4]

1.2 Growth Methods of 2D TMDs

The large-scale growth of semiconducting thin films forms the basis of modern electronics and optoelectronics. A decrease in film thickness to the ultimate limit of the atomic, sub-nanometer length scale, a difficult limit for traditional semiconductors (such as Si and GaAs), would bring a wide range of applications in flexible monolayer electronics, photovoltaics, and display technology. Synthesizing TMDs with specific properties to fulfill the aforementioned applications is crucial to ensure the best performance. To produce TMDs, two approaches are followed; (i) Top-down which mainly relies on physical exfoliation.[5] (ii) Bottom-up which can be achieved by vapor deposition (chemical or physical),[6] or solution-based synthesis.[2] For most of the applications, high quality, and large-area TMDs are a necessity to improve the yield. To attain such requirements, chemical vapor deposition (CVD) has been the most used growth method.

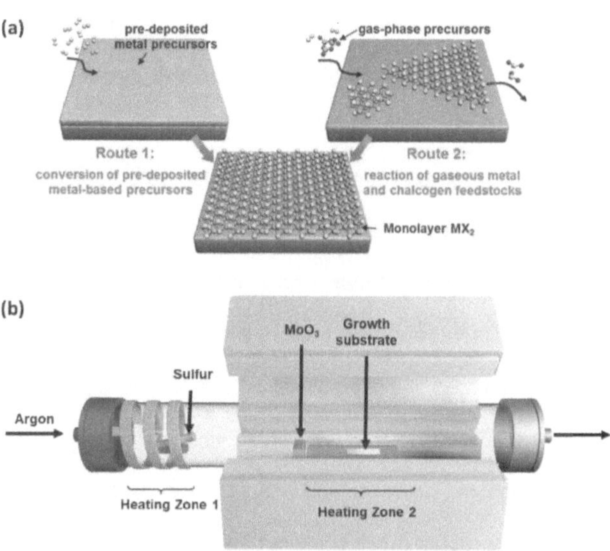

Figure 1.2 CVD synthesis of MX$_2$ monolayer. (a) MX$_2$ synthesis routes.[7] (b) Typical experimental setup of the MX$_2$ growth.

In CVD growth, one of the following routes is used to synthesize monolayer MoS$_2$ flakes as shown in Figure 1.2. (i) route 1: depositing of Mo-based precursors on a substrate and then converting the Mo precursor film into MoS$_2$ via sulfurization.; (ii) route 2: evaporating solid-phase precursors that proceed with gas-phase reactions before depositing on the substrate of interest.[7] The direct chemical vapor phase reaction of MoO$_3$ and S powders route has shown thickness control of MoS$_2$ over a large-area reported by Li's group.[8, 9] Specifically, this method enables the direct growth of single-crystal MoS$_2$ domains with higher quality on arbitrary substrates by controlling the nucleation density and growth parameters. A wide variety of

TMDs (WS_2, WSe_2, $MoSe_2$, *etc.*) has been synthesized using either of the aforementioned routes. Figure 1.2 schematically depicts a typical experimental setup of the CVD synthesis of MX_2. In a furnace equipped with a quartz tube, the transition metal oxide powder is placed in a ceramic boat located in heating zone 2 in the center of the furnace. The chalcogenides powder is placed in a separate quartz boat at the upper stream side and heated using a heating belt in heating zone 1. The reactants are delivered to the substrate using a carrier gas and the reaction is maintained at low pressure via a mechanical pump. CVD synthesis has been the workhorse that consistently delivers a quality synthesis of TMDs in a controllable fashion. However, the dearth of well-elucidated growth mechanisms has led to the produced materials with limited spatial uniformity and electrical performance. Particularly, the electrical performance of the resulting materials, which is often reported from a small number of devices in selective areas, fails to deliver the spatially uniform high carrier mobility.

1.3 CVD Growth Mechanism of TMDs

In CVD, the reactions can be divided into different phases based on reaction schemes. These include evaporation, deposition, nucleation, and diffusion on the substrate as shown in Figure 1.3. Nucleation and diffusion are the most important steps to define the orientation and the shape of the resultant domains. Generally, controlling these steps confers more freedom for 2D materials growth with special characteristics. Different scenarios of TMDs nucleation and growth have been reported that are generally attributed to the wide range of growth conditions within the CVD systems, and interactions with the underlying substrate. However, there is yet to be a universal mechanism that fits all of the reported scenarios. Therefore, comprehensive elucidation on the nucleation and growth mechanisms shall point to the path toward the scalable growth of electronic grade TMDs.

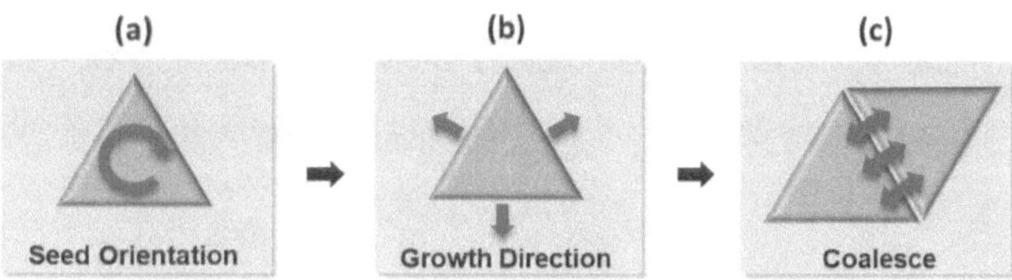

Figure 1.3 Schematic illustration of TMDs film growth process. (a) The growth starts with seeding in a preferred orientation. (b) The reactants diffuse on the substrate and attach to the edge of the seed to extend the domain. (c) Large domains merge to form a continuous film.

1.3.1 Nucleation

In general, vapor phase reactions in CVD growth involve two regimes; (i) homogenous reactions where the reaction takes place in the gas phase under high temperature and partial pressure, followed by condensation directly on the substrate, or (ii) heterogeneous reactions where the reaction occurs on the substrate's surface. Suboxide particles are deposited on the substrate and then interact with chalcogens as shown in Figure 1.4a.[10] As most of the CVD growth includes both regimes, the former should be avoided which adversely causes particle deposition and defect formation. For that reason, most CVD growth of 2D

TMDs proceeds under low pressure. To understand the nucleation in the heterogeneous reaction, Cain et al. used aberration-corrected scanning transmission electron microscopy (STEM) to determine the structure of the formed seed. Figure 1.4c-e shows a nucleation center with a core/shell configuration that contains an oxygen-rich core, typified by the formation of $MoO_{3-x}(S, Se)_y$ nanoparticles in a chalcogen-deficient atmosphere are served as seeding for the TMDs, i.e., the TMDs growth follows a "self-seeding" process.[11] On the other hand, the seeds in the chalcogen-rich atmosphere can be consisted of pure MoS_2 layers without oxygen as we will describe in Chapter 3.

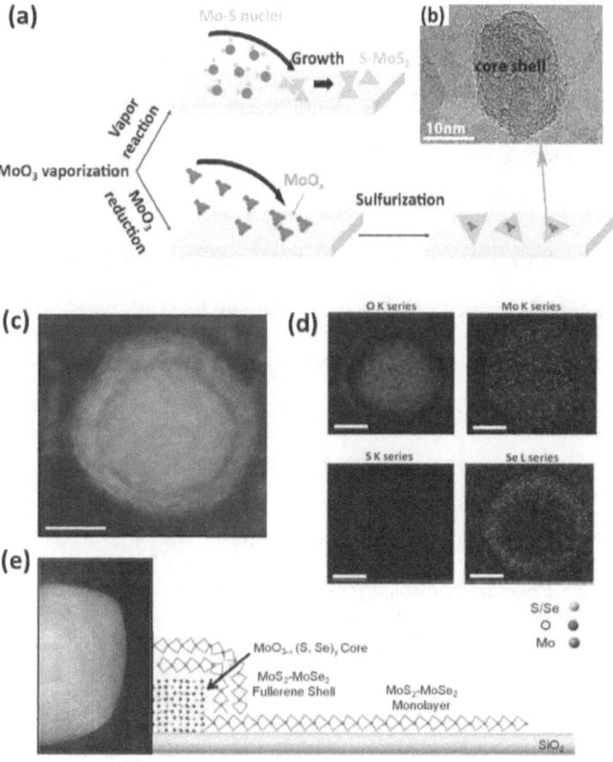

Figure 1.4 TMDs nucleation in CVD growth. (a) Schematic illustration of the possible nucleation scenarios. (b) High-Resolution Transmission Electron Microscopy (HRTEM) shows the core-shell structure for the MoO_{3-x} seed.[10] (c) High-magnification High-Angle Annular Dark-Field (HAADF) image of the MoS_2-$MoSe_2$ nucleation center. (d) The corresponding energy-dispersive spectroscopy (EDS) maps of O, Mo, S, and Se shows that the nucleation center contains oxide. (e) Schematic Cross-sectional of the nucleation centers with the oxide-containing core.[11]

1.3.2 Diffusion

Before the reactants reach the substrate surface, the growth temperature determines the growth behavior whether is thermodynamically or kinetically controlled. Subsequently, the diffusion of the reactants will be guided by the substrate features, like steps, or the substrate-lattice where the reactants can have a specific preferred direction on asymmetrical substrates which results in an unusual domain morphology. Figure 1.5 shows how the substrate-lattice defines the morphology of the domains on an asymmetrical substrate such as a-plane sapphire and $SrTiO_3$. In this study, the effect of the preferred diffusion path on the nanoribbons growth will be described in Chapter 4.

Nucleation and diffusion are the critical steps to start the growth and define the resultant film through controlling the orientation of the seeds and guiding the growth isotropy. All the steps are influenced by the growth process parameters such as temperature, pressure, precursor ratio, flow rate, and choice of substrate. All of which can be controlled to achieve specific morphologies and properties. The mechanism of the resultant growth differs in term of whether TMDs are nucleated from TMDs seeds or suboxide particles. The effect of the different seeding process on the epitaxial growth will be described in Chapter 3. The exact growth mechanism remains largely unexplored and is highly affected by the conditions and the substrate. The growth can be started and guided by either the substrate lattice, steps, ledges, defects, or interfacial layer that is formed on the substrate during the growth. Promoting and guiding the growth using external particles or patterning the substrate are also widely used techniques to control the orientation and morphology of the resultant domains. Each of those growth processes has a different mechanism and has to be precisely studied in a specific environment to achieve full control of TMDs growth as will be discussed in the following sections.

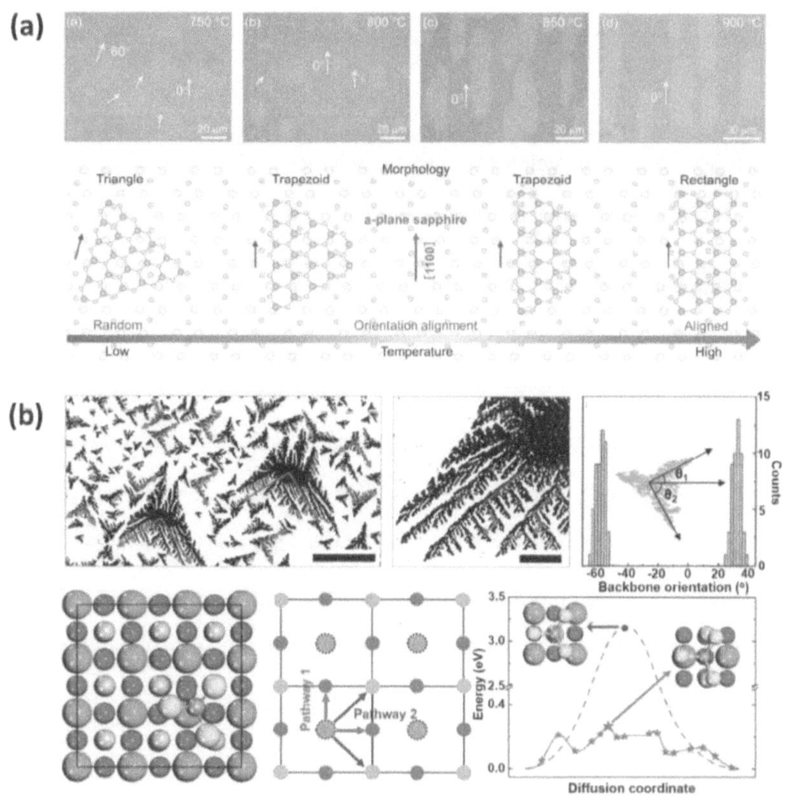

Figure 1.5 Effect of the substrate-lattice (asymmetrical crystal) on the growth morphology. (a) growth of TMDs on a-plane sapphire results in rectangular domains.[12] (b) Growth of TMDs on SrTiO3 results in dendritic morphology.[13]

1.3.3 Growth modes

In practice, the interaction between film and substrate plays a very important role in determining the initial nucleation and film growth. Many experimental observations have concluded that there are three basic nucleation modes as schematically represented in Figure 1.6:

(i)   Layer-by-Layer (Frank-van der Merwe) growth mode

(ii)   Island (Volmer-Weber) growth mode

(iii)   Layer-Island (Stranski-Krastanov) growth mode

Epitaxial growth of TMDs can follow one of the aforementioned growth modes depending on the growth conditions and environment. Monolayer TMDs growth usually adopts the Layer-Island growth mode. In this mode, the different interfaces *i.e.,* between substrate/TMDs and TMDs/TMDs are the major driving force toward different growth modes in the first layer where the domains are grown and then coalesced to form the first monolayer then the rest of the layers are grown as islands. On the other hand, multilayer island TMDs are present in Volmer-Weber growth mode with abundant structural edges which is beneficial for catalysis applications.[14]

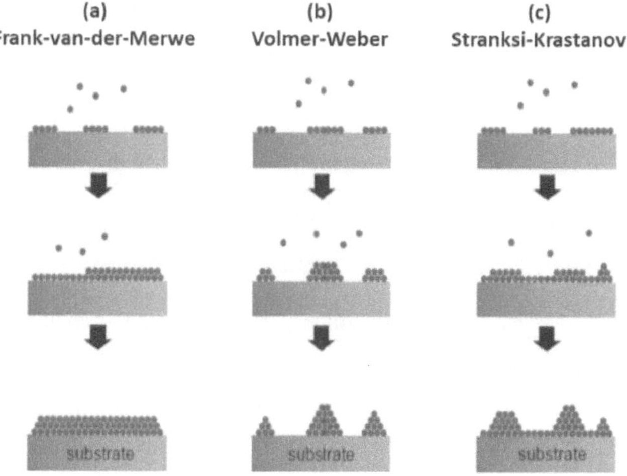

Figure 1.6 Schematic illustration of the three different growth modes. (a) Frank-van-der-Merwe or layer-by-layer growth (b), Volmer-Weber or island growth (c), and Stranski-Krastanov or Layer-island growth.[15]

1.4 Controllable Growth of High-quality Large-scale Single-crystal and Defined Morphology TMDs

Synthesizing high-quality 2D TMDs is essential to realizing the vision of monolayer electronics and optoelectronics. Several factors have been shown to affect the quality and will need to be avoided during the synthesis process. First, the impurity of the precursors can lead to contamination of TMDs film and forming a mid-gap level which has been confirmed by many studies.[16-18] The sources of the contamination could be anything ranging from the system parts to substrate holders, precursors, or the substrate, especially when treated chemically or electrochemically, making it difficult to deliver consistent and reproducible growth of high-quality 2D TMDs. Using highly purified precursors, i.e., with less foreign materials, has been shown to produce higher quality materials as is evidenced by electrical and optical characterizations. Treating the substrate or adding a film onto it were proven to help promote the growth nucleation, assist the lateral growth, and control the orientation of the domains. However, these materials affected the quality of TMDs. Using a chemically inert substrate can be useful to get a contamination-free film.

Besides, grain boundaries are present when random domains are merged to form a continuous monolayer. These structural discontinuities can act as a sink of the electronic carriers and hence degrade the mobility and associated electrical properties.[19] Therefore, growing grain boundary-free monolayer has been the top priority of 2D materials research communities. There are two possible strategies to grow a single-crystal 2D monolayer film. The first strategy is decreasing the nucleation density to grow one large grain. Recently, different materials with a large grain monolayer can be grown such as MoS_2,[20] WS_2,[21] WSe_2,[22] and $MoTe_2$.[23] However, this strategy may not be practically reproducible and not generally applicable to achieve wafer-scale film. On the other hand, epitaxial growth on a crystalline substrate can produce an epitaxial film by merging unidirectional domains. Unlike TMDs, graphene's isotropic structure allows the domains to coalesce without mirror grain boundaries which usually formed as a result of merging inverse domains. Wafer-scale, grain boundary-free, and monolayer graphene have been achieved by the unidirectional growth where multiple domains are merged on Si wafer with a hydrogen-terminated germanium buffer layer.[24] Most of the substrate used to grow TMDs are symmetrical, resulting in similar registration energy for the 0° and 180° domains which form a mirror grain boundary upon merging into a film. Single-crystal film TMDs growth can be achieved by lifting the 0°/180°stacking degeneracy.[25] Different strategies have been followed to control the TMDs growth for different purposes. In the following sections, we will illustrate some of these strategies.

1.4.1 Precursors-assisted growth

Controlling the precursors' evaporation rate is important to ensure proper nucleation density to enlarge the domain size. Adding salts to the metal oxides has been tested in a myriad of 2D TMDs (Figure 1.7a), and proved to decrease the metal oxides melting point. Similar to the growth of 1D nanowires, salt-assisted epitaxy growth takes advantage of the decreased melting point that ensures the formation of more intermediate products and hence increases the reaction rate. Besides facilitating the nucleation on the substrate, larger domains are achieved under the fast reaction rate by following this approach.[26] Using salts has been also shown to control the morphology of the resultant growth. For example, nanoribbons have been achieved using salt mixed with the metal oxide.[27] In this scenario, droplets containing salt mixed with the metal oxide would form and crawl on the substrate, forming TMDs along the deposition line with high crystallinity and aspect ratio as shown in Figure 1.7 b.

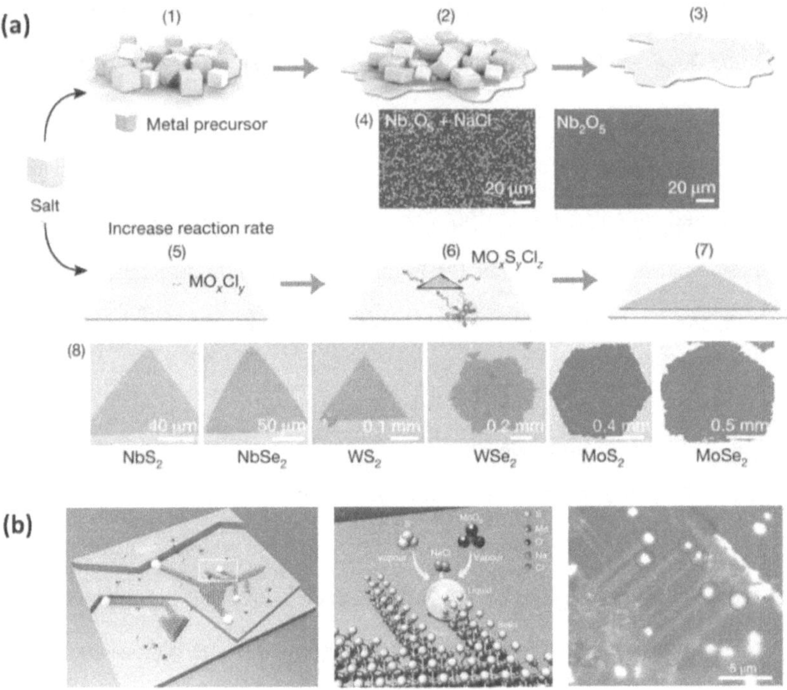

Figure 1.7 Salt-assisted growth. (a) Schematics of the salt-assisted growth process (1-3). (4)Scanning Electron Microscopy (SEM) images of the Nb seeds with (left) and without (right) added salt.[26] (b) Growth of nanoribbons guided by Na-Mo-O droplets by vapor-liquid-solid growth.[27]

1.4.2 Gas-assisted growth

Controlling the gas-phase reaction is essential to determine the structure of the seeds which leads to different interactions with the substrate. Besides using inert gases to deliver the reactants to the substrate, other gases can be introduced to achieve many purposes such as enlarge the domain size, growing heterostructures, or selective growth following substrate features. Large single domain MoS$_2$ up to 350 µm has been achieved by introducing a small amount of O$_2$ to the growth.[28] The etching effect of oxygen is an important factor to reduce the nucleation density and make room to enlarge the single domain size. However, the high etching process can decrease the size or eliminate growth. To this end, balancing between the growth and etching is crucial as shown in Figure 1.8.[28] Moreover, oxygen can prevent poisoning the precursors and ensure a continuous supplementation of the precursors through the growth time and results in a large-area TMDs. Yu and co-workers have successfully synthesized a wafer-scale continuous monolayer MoS$_2$. By isolating the MoO$_3$ precursor from S, not only prevents the metal oxide from sulfurization and forming unwanted compounds but also protects it from consumption during the growth. The interaction between the grown film and sapphire substrate was mostly vdW epitaxial interaction and no interlayer bonds with sapphire are observed. Subsequently, the grown film can be easily transferred from sapphire onto a target substrate and the sapphire reused for multiple growths.[29]

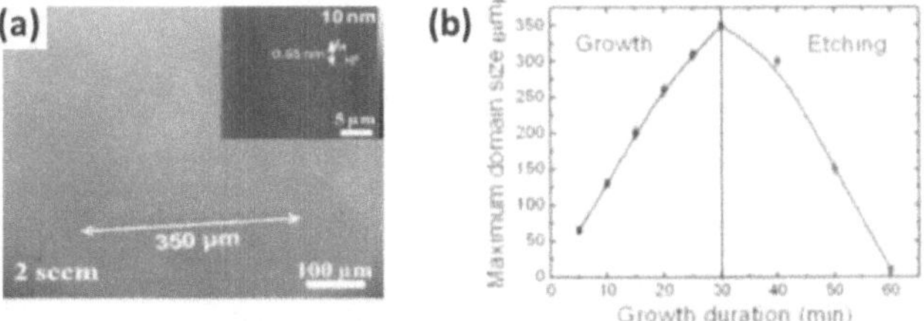

Figure 1.8 Effect of the oxygen on the domain size. (a) Optical image (OM) of large domain MoS$_2$ (b) The evolution of the domain size as a function of growth time in oxygen environment.[28]

Hydrogen is another example of carrier gases that affects the nucleation and growth of TMDs. Introduction of H_2 has two different roles; (i) reducing the metal oxides and chalcogenide precursors more effectively which determine the nucleation and the quality of the film as shown in Figure 1.9a-f [30], and (ii) passivating the substrate which affects the surface diffusion of the precursors resulting in different domain shapes, sizes, and thickness.[31] The latter has an important role in selective growth where the nucleation can be modulated by controlling the H_2 gas concentration and introduction time. The substrate terminated with H_2 can determine whether seeds form between the steps, or on the step-edges resulting in two different growth mechanisms; step-edge aligned growth (Figure 1.9g), or substrate-lattice aligned growth (Figure 1.9i).

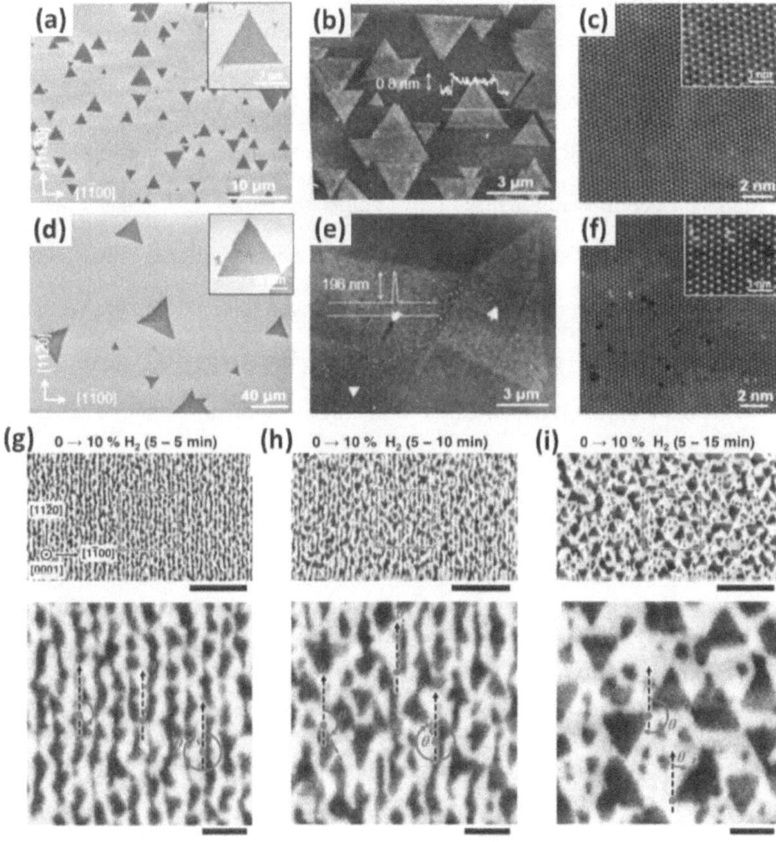

Figure 1.9 Hydrogen assisted growth. (a-f) The effect on the growth quality with (a-c) and without (d-f) introducing H_2.[30] Changing of the growth mode from step-directed (g) to lattice-directed (i) growth with different H_2 duration.[31]

1.4.3 Template-guided growth

Epitaxial growth of single-crystalline material on a single-crystalline substrate that has a lattice matching with the lattice of grown material is called "epitaxy". However, the weak vdW interaction in the epitaxial growth of layered materials on dangling bonds-free substrate (vdW epitaxial) (Figure 1.10) allows the remaining strain, as a result of lattice mismatch, to be accommodated in the vdW gap and hence layered materials can be grown even with large lattice mismatch.[32] Therefore, a wide range of substrates with different lattice structures can be used. In the pioneer reports, a substrate with only hexagonally-arranged lattices such as graphene,[33, 34] sapphire,[29, 35, 36] mica,[37] GaN,[38, 39], and h-BN[40] have been widely used to accommodate the hexagonal TMDs lattice as shown in Figure 1.11a-e. Recently, substrates with non-hexagonal lattice structures are used for TMDs growth such as substrates with cubic lattice such as Au[41, 42] and SrTiO$_3$[13] as shown in Figure 1.11f,g.

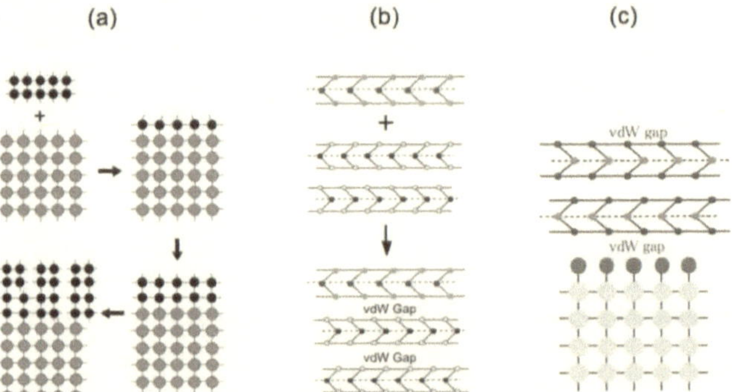

Figure 1.10 Epitaxial growth type. (a) Conventional epitaxy. (b) vdW epitaxy. (c) heteroepitaxy. [32]

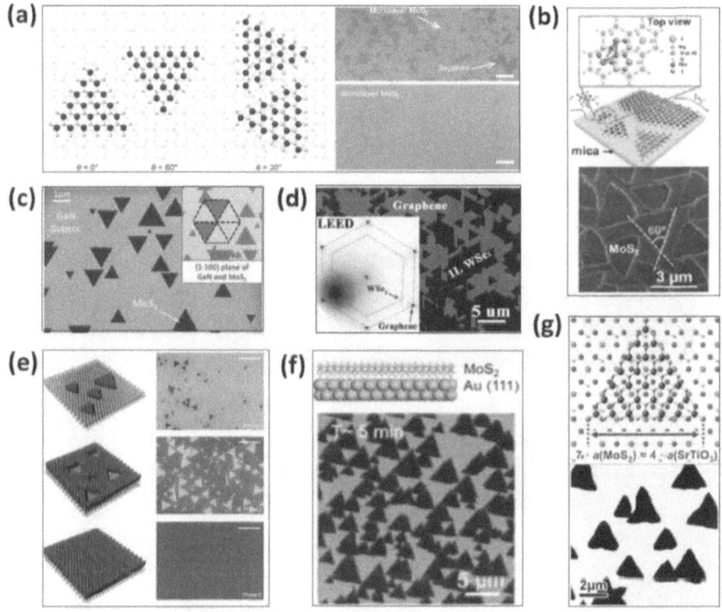

Figure 1.11 Epitaxial growth of TMDs on different substrates. (a) Sapphire.[35] (b) Mica.[37] (c) GaN.[39] (d) Graphene.[33] (e) h-BN.[40] (f) Au[41, 42]. (g) SrTiO₃.[13]

The potential of the interaction between the TMDs and the substrate is not limited to the interaction through the vdW gap, but also can be screened even with the existence of the graphene layer on top of the substrate.[43] This approach has offered the possibility to grow, transfer, and stack different 2D materials without any dangling bonds that can degrade the quality of 2D materials. It has been suggested that the polarity of the substrate and the 2D materials governs the transparency of the 2D materials to remote epitaxial growth which is also affected by the number of grown 2D materials layers as shown in Figure 1.12.[44]

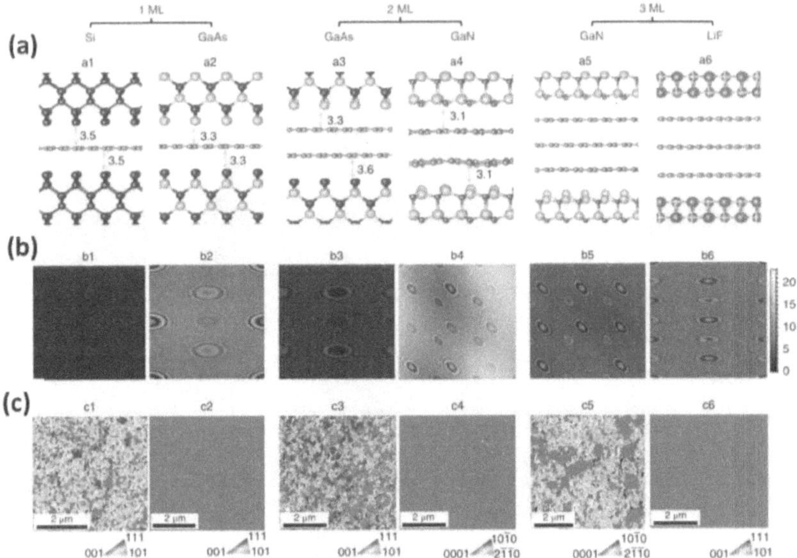

Figure 1.12 Remote epitaxy. (a) Atomic structure of Si, GaAs, GaN, and LiF on 1-ML-, 2-ML (bilayer)- or 3-ML (trilayer)-graphene-coated Si, GaAs, GaN, and LiF, respectively. C, grey; Si, dark blue; Ga, light green; As, purple; N, blue; Li, dark green; F, orange. (b) Maps of potential fluctuation. (c) Electron backscatter diffraction (EBSD) map of released surfaces.[44]

Selective growth is one of the important demands for future electronics to achieve specific requirements such as device-ready architecture to avoid the damage associated with the post-processing the film. For example, lateral or vertical heterojunctions growth where two dissimilar materials are grown,[45, 46] nanoribbons, and mono-nucleation for single-crystal film growth have shown improvement in the quality of 2D materials. Generally, two approaches have been used for selective growth: pre-patterning the substrate before the growth, or using naturally occurring features on the substrate to act as a template for guided-growth. While the former approach is the more controllable process, the latter has the important advantage of limiting the contamination associated with the patterned materials.

Patterning is the straight forward method to selectively grow TMDs following a pre-designed template. Recently, seed-promoted growths have been realized on pre-patterned substrates using photolithography and O_2 plasma.[47] Plasma treated areas on the SiO_2 substrate have higher surface energy, which promotes selective growth. The patterned substrate can be utilized repeatedly with well-controlled growths and transfers as shown in Figure 1.13a. Selective patterning of monolayer or bilayer TMDs using a laser in a controllable and precise fashion to create nucleation sites allowed Li *et. al.* realize a wide range of vdW heterostructures arrays (Figure 1.13b), including VSe_2/WSe_2, $NiTe_2/WSe_2$, $CoTe_2/WSe_2$, $NbTe_2$/WSe_2, VS_2/WSe_2, VSe_2/MoS_2, and VSe_2/WS_2.[46]

Both previous approaches are based on creating sites with higher surface energy to promote TMDs growth. Metal oxides also can be pre-patterned on the substrate to grow TMDs in a selective area. In recent work, the metal oxides (WO_2) and (MoO_2) were patterned on a substrate to grow p- and n-type TMDs selectively as shown in Figure 1.13c.[49] This metal-guided selective growth technique can be used to realize the wafer-scale and location-selective growth of two dissimilar TMDs grown simultaneously in a one-step CVD process. Alternatively, the TMD itself can be patterned using a stamping method. Dense arrays of MoS_2 nanoribbons with high aspect ratio can be patterned using a stamp printing and thermolysis process with thiosalts used as ink patterned on an arbitrary substrate as shown in Figure 1.13d.[48] However, monolayer and few layers can't be achieved

using this strategy.

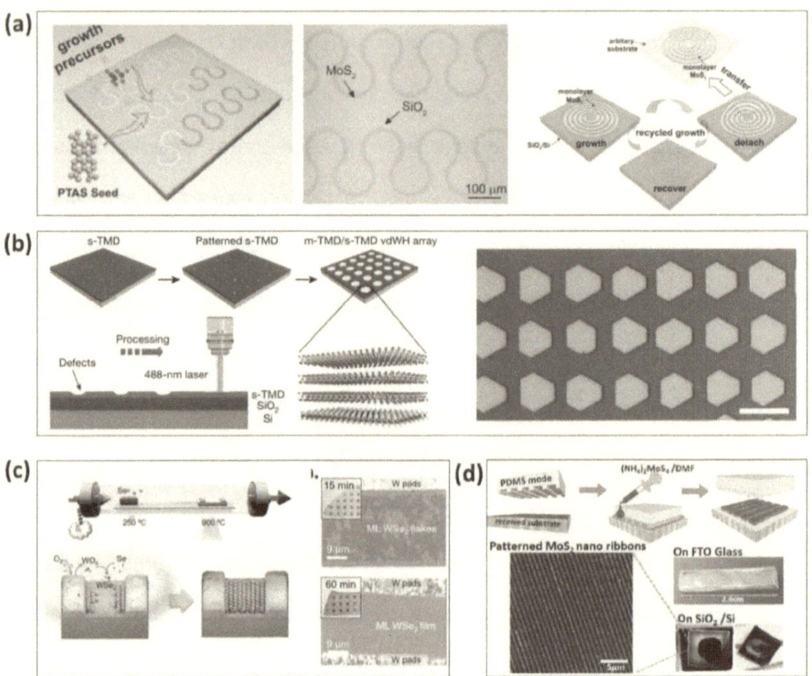

Figure 1.13 Selective growth for controllable morphology. (a) Patterned growth of TMDs via seed-promoted growth by lithography patterning and O_2 plasma etching.[47] (b) Selective patterning on the monolayer TMDs using a laser to create defects acting as nucleation sites for a second growth.[46] (c) Metal-guided selective growth of TMDs films.[49] (d) Patterning dense arrays of MoS_2/MoS_x nanoribbons by a stamp printing and thermolysis process with thiosalts as ink.[48]

Besides patterning, using a naturally accruing template ensures a high quality of the grown materials without the contamination associated with patterned materials. Step-edge guided growth has been realized firstly in 2015 where WSe_2 was grown on c-plane sapphire.[50] It has been found that the high growth temperature (950°C) defines the sapphire step edges that facilitate the nucleation at the step-edge. Although unidirectional nucleation can be achieved in this strategy as shown in Figure 1.14a, the isolated domains are grown on top of each other upon enlarging the domain size due to the high steps and hence growing large-area monolayer film was not possible. The determining factors to allow either substrate lattice-guided growth or substrate steps-guided growth were not clear. Recently, however, a systematic study described how introducing H_2 gas plays a crucial role in controlling the growth guidance whether the growth is step-edge guided (Figure 1.9g), or substrate-lattice guided (Figure 1.9i).[31]

Copper (Cu) is another example of a substrate used to grow 2D materials such as graphene and h-BN. The growth of single-crystal graphene film can be easily achieved on any symmetrical substrate due to the compatibility with the 6-fold symmetry of graphene. On the other hand, 2D materials with 3-fold symmetry need more effort the break the symmetry between the antiparallel domains by directly using a substrate with asymmetrical crystal structure, or pronounced step edges to guide the growth unidirectionally. The coupling between copper step edges and h-BN zigzag step allows the growth of centimeter-scale and high-quality h-BN film on the low-symmetry Cu (110) Figure 1.14b.[51] While the growth on the symmetrical facet like Cu (111) was known to grow antiparallel domains of h-BN due to the 3-

fold symmetry of this material, annealing the Cu (111) to control the steps and hence break the symmetry allowing the growth of single-crystal h-BN film as shown in Figure 1.14c.[52] However, the Cu substrate is not suitable for 2D TMDs growth due to the interaction between the chalcogens and the copper. Recently, single-crystal MoS_2 film was achieved by inducing unidirectional nucleation along Au (111) step edges as shown in Figure 1.14d.[42] More efforts are needed to apply those methods on a wafer-scale and produce high-quality materials for practical applications.

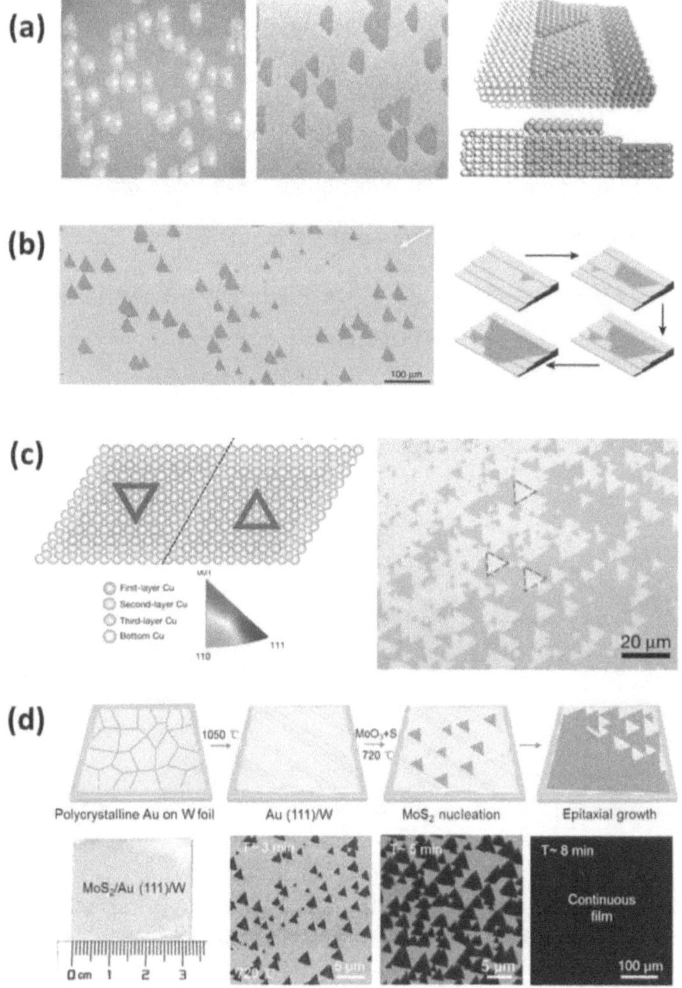

Figure 1.14 Step- guided growth. (a) TMDs on sapphire.[50] (b) h-BN on Cu (110).[51] (c) h-BN on Cu (111).[52] (d) TMDs on Au (111).[42]

Besides utilizing the substrate steps to break the symmetry of the symmetrical substrate, the steps can guide the growth in a specific direction and form nanoribbons. More specifically, when the step edge is wide, it exposes a specific facet called a ledge. Graphene nanoribbons have been grown on SiC ledges (Figure 1.15) after annealing the substrate at 1,200–1,300°C to induce (1-10n) facet, whereas further heating to 1,450°C forms multilayer graphene nanoribbons. However, this method is limited to graphene nanoribbons. In this study, a substrate with wide ledges such as β-gallium (III) oxide (β-Ga_2O_3) (100) was used as a template for nanoribbon growth. The synthesis process will be described in Chapter 4.

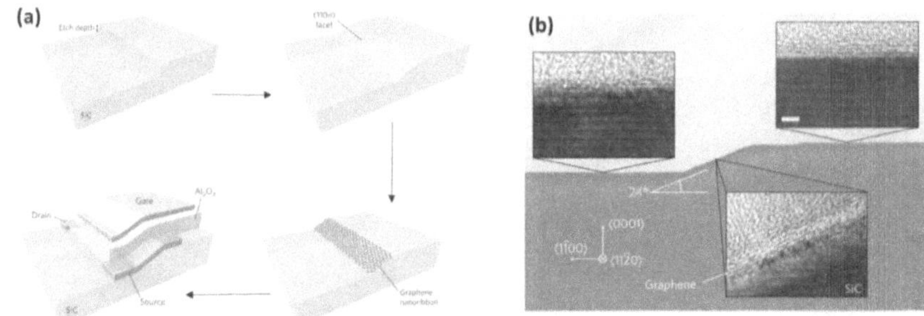

Figure 1.15 Tailoring graphene nanoribbons on SiC. (a) Schematic illustration for the growth process and device fabrication. (b) Cross-sectional HRTEM of graphene selectively grown on the step.[53]

1.5 Objectives and Contributions

Epitaxial single-crystal growth is an important approach to synthesize grain boundary-free monolayer 2D TMDs. The high-quality film ensures the full potential of these materials is utilized to alternate the traditional silicon-based electronics. However, epitaxial growth is affected by several factors that are hard to control at the one time due to the lack of a full understanding of the CVD growth mechanism. These factors can be controlled in two different phases; the gas-phase (gas assisted growth), and the solid-phase (template-assisted growth). Combining these factors is essential to control the domains' orientation and morphology. As a result, single-crystal film, or arrays of nanoribbons can be achieved in a controllable fashion. This book reports new growth methods which allow controlling the orientation and the dimensionality of the MoS_2 as an example of 2D TMDs. Additionally, this book discusses the artifact arises during the imaging of 2D TMDs which pronounce along the grain boundaries.

In Chapter 1, the study background and book objectives are presented. In Chapter 2, we review the experimental methods that were essential in synthesizing, characterizing, and evaluating the overall quality of our 2D materials. In Chapter 3, we investigate the effect of the precursors' ratio on the composition of the formed seeds and accordingly control the orientation on c-plane sapphire. In Chapter 3, we describe a newly developed method, called Ledge-directed epitaxy (LDE), which allows the growth of single-crystal nanoribbons and fabricate field-effect transistors based on nanoribbons. In Chapter 5, we investigate the effect that appears during grain boundaries imaging using STEM that can lead to misinterpretation of the 2D TMDs phases. Finally, Chapter 6 summarizes the main conclusion and suggestions for future work.

Chapter 2: Sample Preparation and Characterization Techniques

2.1 Sample preparation

2.1.1 Growth of MoS_2 monolayer on sapphire

In this study, the MoS_2 monolayer is grown by the chemical vapor deposition method. The precursors are MoO_3 (Sigma-Aldrich, $\geq$ 99.5% purity) and S (Sigma-Aldrich, $\geq$99.5% purity) powders. The sulfur powder (4 g) is put at the upper stream side of the furnace (heating zone 1), and the temperature is maintained at 140°C during the reaction. In heating zone 2, the center of the reaction chamber, the MoO_3 powder (0.8 g) is placed in a ceramic boat with a 1 cm × 5 cm sapphire substrate placed downstream of the ceramic boat. The gas flow is from Ar (Ar = 70 sccm) and the chamber pressure is kept at 40 Torr. First, the center of the furnace is gradually heated from room temperature to 750°C at a ramping rate of 25°C/min and kept at this temperature for 5 min. Then, the temperature was increased to the growth temperature, 800°C, with the same ramping rate and kept for 10 min. The furnace is then naturally cooled to room temperature.

2.1.2 Growth of MoS_2 and WSe_2 nanoribbons

Single-crystal MoS_2 and WSe_2 monolayer nanoribbons were grown on the β-Ga_2O_3 (100) substrate by the conventional chemical vapor deposition (CVD) in a horizontal hot-wall 2" furnace tube with two heating zones. High purity S (Sigma-

Aldrich, 99.99%), Se (Sigma-Aldrich, 99.99%), MoO_3 (Sigma-Aldrich, 99.9%), and WO_3 (Sigma-Aldrich, 99.9%) powders were used as the reaction precursors. MoO_3 (WO_3) powder typically was placed in a ceramic boat and was put in the heating zone center of the furnace. S (Se) powder was placed in a separate quartz boat at the upper stream side maintained at 140°C (270°C) during the reaction. The single-crystal β-Ga_2O_3 (100) substrate was placed at the downstream side, where the precursor vapors were brought to the substrates by Ar gas flowing at 30 torr for MoS_2 and Ar/H_2 mixture gas at 10 torr for WSe_2. The center heating zone was heated to 800°C and kept for 10 min for the growth of MoS_2 nanoribbons. On the other hand, for the growth of WSe_2 nanoribbons, the furnace was heated to 900°C and was held for 15 min. Upon completion of growth, the furnace was naturally cooled down to room temperature.

2.1.3 Transfer of MoS_2

MoS_2 was transferred onto other substrates using polydimethylsiloxane (PDMS). In brief, a thin PDMS film was placed on top of MoS_2/substrate. After making a conformal contact with the substrate, the sample was soaked in 1M KOH for 5 min at room temperature. After rinsing the sample with DI water, the PDMS with MoS_2 was slowly peeled-off and placed on the target substrate. The sample was kept in a vacuum for 30 min to increase the adhesion of MoS_2 on the substrate, and dry out any residual water drops. Finally, the PDMS was peeled off leaving the MoS_2 on the target substrate.

2.2 Optical Characterization

Raman spectroscopy and photoluminescence are the most widely used tools in the study of 2D crystals. Raman spectroscopy is used to determine the crystallinity and number of layers while photoluminescence is used to determine the bandgap energy.[54] Electronic modification of MoS_2 monolayer can be demonstrated via studying the change in the PL and Raman peaks which can be a consequence of strain, defects, or doping from the substrate.

2.2.1 Strain

Raman spectroscopy was adopted by Hairam, *et al.* to observe phonon softening with increased strain, breaking the degeneracy in the E' Raman mode of MoS_2. Moreover, by using photoluminescence spectroscopy, they measured a decrease in the optical band gap of MoS_2 that is approximately linear with strain ~45 meV/% strain for monolayer MoS_2. Also, they observed a pronounced strain-induced decrease in the photoluminescence intensity of monolayer MoS_2 that is indicative of the direct-to-indirect transition of the character of the optical bandgap of this material at an applied strain of ~1%.[55] Evolution of the Raman spectrum and PL as a device is strained from 0 to 1.6% are shown in Figure 2.1. By using these indicators, any strain induced in the MoS_2 monolayer grown on sapphire or other substrates can be determined.

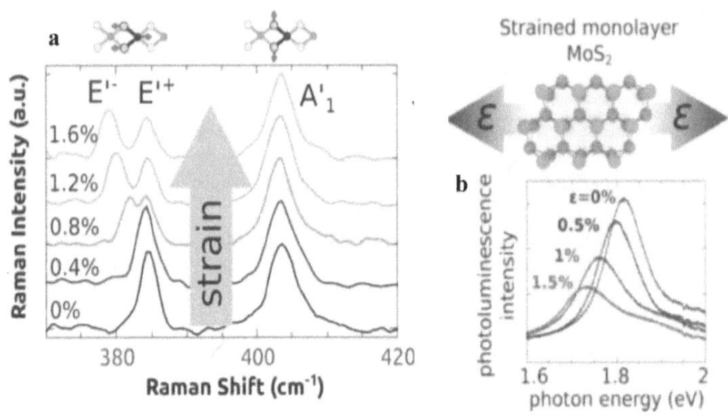

Figure 2.1 Evolution of the Raman and PL spectrum. Evolution of Raman (a) and PL (b) as a device is strained from 0 to 1.6%.[55]

2.2.2 Doping

To determine the doping density, Kin, *et al.* fabricated field-effect transistors (FETs) using MoS_2 on SiO_2/Si substrates. The doping density in the MoS_2 channel was systematically varied by applying a voltage to the Si back gate. The photoluminescence intensity of the A exciton, like its absorbance, can be switched off by doping. The absorption and especially photoluminescence are highly dependent on doping. In addition, redshifts in the photoluminescence peak energies from the corresponding absorption energies (Stokes shifts) are observed for both features. The magnitude of the Stokes shift was found to increase with the doping level as shown in Figure 2.2a-b.[56] Electron doping in single-layer MoS_2 results also in softening specifically of its Raman-active A_{1g} phonon, accompanied by an increase in the linewidth of its Raman peak. In comparison, the other Raman mode with E^1_{2g} symmetry is quite insensitive to electron doping. This is due to a stronger electron-phonon coupling of the A_{1g} mode than of the E^1_{2g} mode as shown in Figure 1.17c-d.[57] These variations in PL and Raman peaks allow us to determine any change in the doping level.

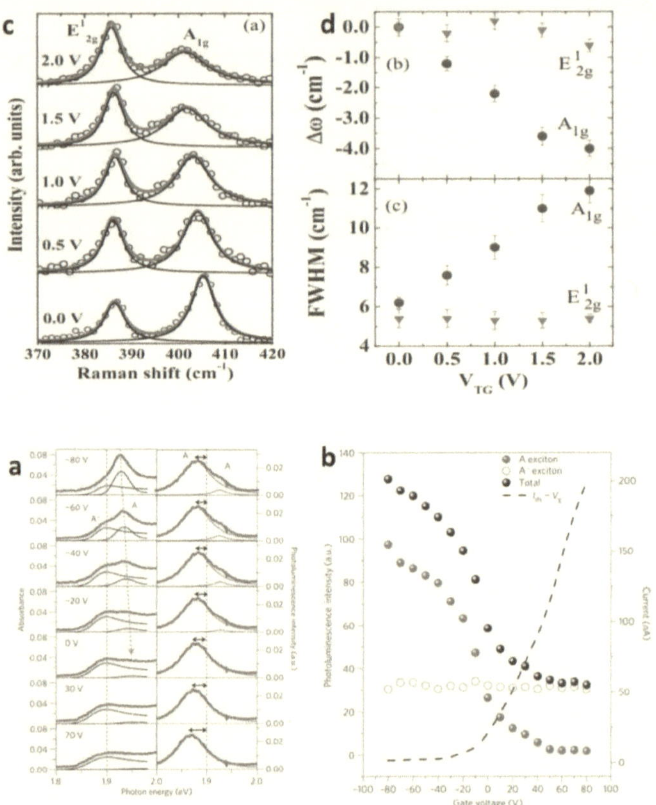

Figure 2.2 Doping dependence of the optical properties of a monolayer MoS₂ FET. (a) Absorption and photoluminescence spectra (red lines). (b) Dependence on gate voltage of the drain-source current (right) and the integrated photoluminescence intensity of the A and A⁻ features and their total contribution (left).[56] (c) Raman spectra of monolayer MoS₂ at different top-gate voltages V_TG. (d) Change in the phonon frequency ω and FWHM of A_{1g} and E^1_{2g} modes as a function of V_TG. [57]

2.3 Grain Boundaries (GBs) Visualization

2.3.1 Sulfurization and oxidization

There are several ways to visualize the grain boundaries. Sulfurization and/or oxidization are the easiest methods due to the high reactivity of GBs with regard to oxidation or impurity adsorption in an ambient atmosphere. In experiments involving mild oxidation under moisture-rich conditions, the GBs of intersecting or adjoining flakes (Figure 2.3a-d) can be readily visualized by scanning electron microscopy (SEM) as bright-line shaped contrasts as shown in Figure 2.3e-h.[36]

2.3.2 Second harmonic generation (SHG) measurements

Recently, a new characterization technique known as SHG has been introduced for identifying GBs embedded in continuous MoS_2 films (Figure 2.3i–l).[58] This allows the few-atom-wide line defects that stitch together different crystal grains to be visualized through destructive interference and annihilation of second-order non-linear waves from neighboring atomic domains (Figure 2.3j). By analyzing the polarised components of this SHG, the crystal orientation of the polycrystalline membrane can be resolved (Figure 2.3k and l). In effect, this technique allows for the high-throughput mapping of crystal grains and grain boundaries over large areas with non-invasive all-optical operations.

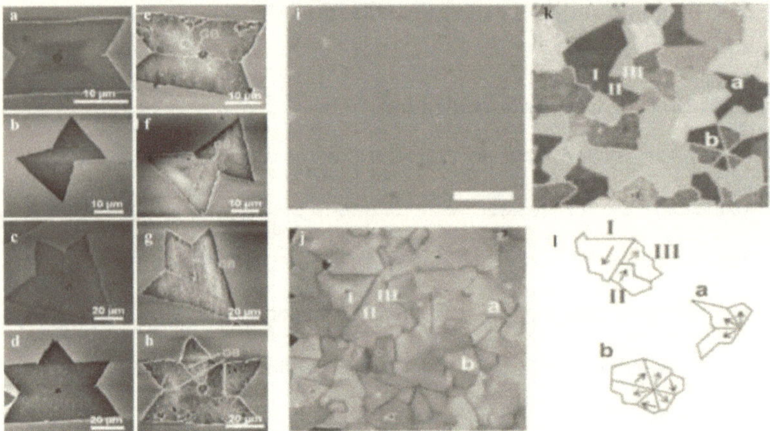

Figure 2.3 Grain boundaries visualization. (a-d) SEM images of as-grown WS$_2$ monolayer polygons. (e-h) SEM images of WS$_2$ grain structures following mild oxidation.[36] (i, j) Optical and SHG images, respectively, of a large-area CVD-grown monolayer of MoS2 on SiO$_2$/Si. (k) SHG image of the same area as that of (i, j) showing the crystal orientations of irregularly shaped polycrystalline aggregates. (l) Schematic of the area marked in (k), with the arrows indicating domain orientations.[58]

2.3.3 High-Resolution Scanning Transmission Electron Microscopy (HR-STEM)

HR-STEM is an advanced imaging technique widely used in imaging 2D materials.

Figure 2.4 shows mirror grain boundaries between two antiphase MoS$_2$ domains.

Besides direct visualizations of GBs, microstructure visualization of 2D materials

allows determining many characteristics such as different phases with different

electrical and optical properties as shown in Figure 2.5.[59]

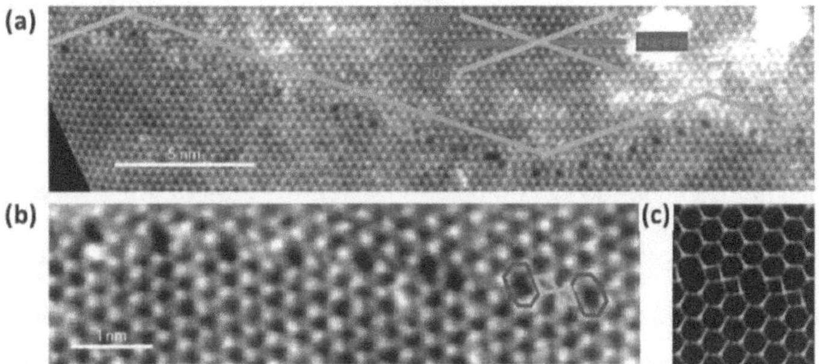

Figure 2.4 Grain boundaries visualization using HR-STEM. (a) mirror GB of MoS₂ monolayer. (b) Zoom-in image of the GB in (a) shows periodic ring defects. (c) atomic model of the experimental structure shown in (b).[60]

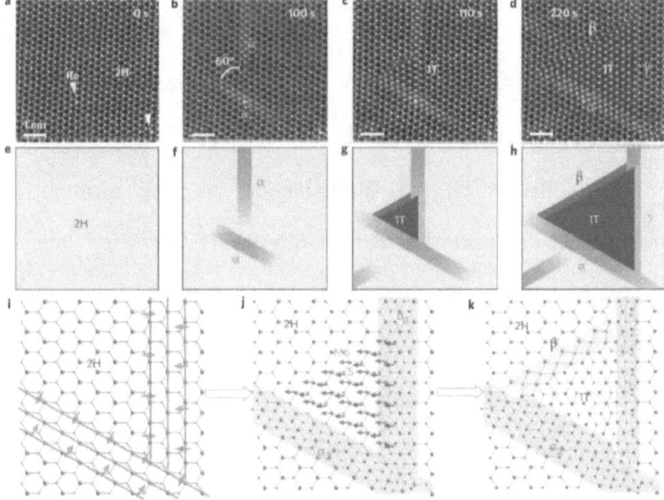

Figure 2.5 Phase transformation of MoS₂ monolayer at high temperature. (a) MoS₂ monolayer doped with Re substitution dopants (indicated by arrowheads) (b,c) Intermediate phases. (d) The area of the transformed phase is enlarged. Three different boundaries (a, b, and g) are found at the three edges between the 1T and 2H phases. (e–h) Schematic illustrations of the phase transition corresponding to the ADF images in a–d, respectively. (i-k) Atomic model of the suggested phase formation mechanisim.[59]

However, some complications are raised from the astigmatisms associated with the correctors, and hence artificial atomic lattice structures to appear in STEM images, and consequently a complete misinterpretation of imaging results. Such misinterpretation led to confusion in different fields of the 2D materials research community. In Chapter 5, an integrated strategy of correlated simulation packages and HR-STEM are utilized to reveal imaging artifacts as a result of the residual 3-fold astigmatism. HR-STEM ADF imaging was performed with a ThermoFisher Titan Themis Z (40-300 kV) TEM equipped with a double Cs (spherical aberration) corrector, a high brightness electron gun (x-FEG), and an electron beam monochromator. To reduce the electron beam sample damage, we chose to operate the microscope at 80 kV and tune it to minimize 3rd order Cs below the detection limit (few μm). However, some residual 3-fold astigmatism of about 100 nm (as measured by the Thermofisher Cs-Probe corrector software with a standard alignment sample) is intentionally left uncorrected. Images were acquired in diffraction mode with the camera length 115 mm corresponding to collection angles of 41 (inner) and 200 (outer) mrad of a Fischione Annular Dark Filed detector. Probe semi-convergence angle was tuned for 30 mrad with the beam current ranging from 50 pA for the regular STEM to 3 pA for the monochromated beam STEM. For the monochromator operation, the method was first implemented in ref.[61] and described in detail in ref. [62]. As a result, the monochromatic STEM was performed with the energy spread in the beam of about 60 meV (defined as the full width at half maximum of the zero-energy loss peak). Gauss high-pass (to reduce contamination effects) and low-pass (to reduce scanning noise) filtering was used to enhance the contrast of the image[63]. Measurements and calibration of optical elements of Cs probe corrector to adjust the A2 astigmatism were performed by standard ThermoFisher software provided with a Cs probe corrector and by the use of a standard alignment sample (Cross Grating Replica 3 mm, AGS106). A Fischione Dual-Axis Tomography Holder (model 2040) was used as this holder allows the specimen to be fully rotated through 360° in the plane orthogonal to the electron beam.

2.3.4 Differential Diffraction Filtered STEM Mapping of the Domain Orientations

Due to the layered configuration that is comprised of a hexagonally packed layer of metal atoms sandwiched between two layered of chalcogen atoms, the lattice of monolayer MoS_2 can be divided into molybdenum (Mo) and sulfur (S) sub-lattices. Consequently, the 6-fold symmetry of the electron beam diffraction pattern is reduced to the 3-fold symmetry of the corresponding sub-lattices. Here, the first-order diffraction spots are divided into two families, namely k_a (green arrows in Figure 2.4b) pointing toward the Mo sub-lattice, and k_b (red arrows in Figure 2.4b) orienting in S sub-lattices. These two families of spots are not equivalent. For the electron beam with an acceleration voltage of 80 kV, k_a spots are ~10% higher in intensity relative to that of K_b. If the specimen is rotated by 180° (60°), spots of both k_a and k_b families are switched over. This phenomenon allows inferring sample orientations through mapping enabled by employing diffraction spots filtering coupled with the STEM imaging in the so-called microprobe mode. In this light, the convergence angle of the electron beam is reduced down to about

0.5mrad, allowing us to obtain convergent beam diffraction patterns without discs overlapping, i.e., all the diffraction spots are clearly separated, closely resembling the case of the parallel electron diffraction. Next, a single k_a spot can be isolated with an objective aperture, and the STEM image is obtained. This creates an image with all the domains in 0° orientation highlighted in blue color. In the second step, the k_b spot is isolated with the same objective aperture, and another STEM image (from the same sample area) is obtained with all the domains in 180° (60°). To enhance the visual contrast between the domains in different orientations, the first image is subtracted from the second images, thus resulting in differential diffraction filtered STEM mapping of the domains orientations toward 0° highlighted in blue and 180° (60°) orientation false-colored in yellow.

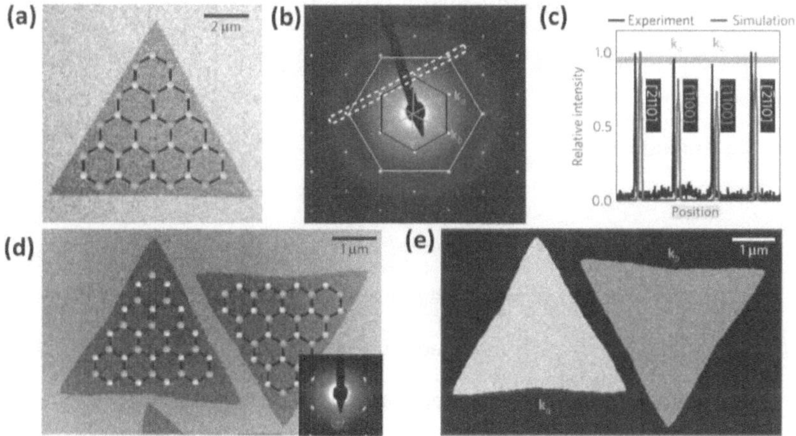

Figure 2.6 Diffraction imaging of crystal orientation and edge terminations.
(a) Bright-field image of a single-crystal triangle with a Mo-zigzag edge orientation.
(b) Diffraction pattern from (a). The asymmetry of the Mo and S sublattices separates the [-1100] diffraction spots into two families. k_a ={(-1100),(10-10),(0-110)} and k_b =$-k_a$. (c) A line profile through experimentally measured diffraction spots (black) and Bloch-wave simulations (red). The higher intensity k_a spots point towards the Mo sublattice, as indicated by the arrows in (a),(b). (d) Bright-field TEM image of two triangles with S-zigzag edge orientations. (e) Dark-field image of the region in (d).[19]

Chapter 3: Substrate Lattice-Guided Seed Formation Controls the Orientation of 2D Transition Metal Dichalcogenides

3.1 Introduction

Two-dimensional (2D) transition metal dichalcogenides (TMDs) have attracted significant attention owing to their unique electrical,[64, 65] optical,[66-68] mechanical[69] and thermal[70] properties inherited from their 2D structures. In clear contrast to the semi-metallic graphene or the unstable black phosphorous,[71] the ambient stable 2D TMDs demonstrate promising properties for electronic and photoelectronic applications, such as field-effect transistors,[64, 72] sensors,[73, 74] solar cells[75] and photodetectors.[66, 76, 77] The molybdenum disulfide (MoS_2) monolayer, with a direct energy gap,[78] strong photoluminescence,[79] efficient valleys and spin control,[65] is one of the most intensively explored member in the 2D TMDs family. It can be obtained by mechanical,[80] chemical[81,82,] or electrochemical[83] exfoliation. However, these strategies lack uniformity and produce defect-rich samples, which may not be suitable for large-scale device fabrication. For this reason, chemical vapor deposition (CVD) methods using the solid precursors MoO_3 and S powders have been developed.[8]

Nevertheless, the developed CVD methods can not guarantee to synthesize defectless TMDs with uniform thickness and orientation. The grain boundary between different domains can lower the quality of TMDs as electronic devices by breaking the structural periodicity and introducing strain defects. Huang *et al.* have further shown that the electronic structures of the grain boundary strongly depends on the misorientation angle of adjacent domains.[16] Hence, recent research efforts have dedicated special attention to control the orientation in order to obtain a large-scale and grain boundary-free monolayer films for higher electron mobility.[84] Owing to the lattice symmetry of the 2D TMDs layers, they are typically grown as triangles, hexagons, or truncated triangles.[85, 86] The growth of aligned TMDs triangles has been achieved on top of other crystalline layered materials such as graphite,[87] graphene[34, 88, 89] or boron nitrides,[90] where the TMDs monolayers are either aligned by the graphite step edges or with the substrate lattice through van der Waals (vdW) interaction. The aligned growth has also been successfully achieved on single-crystal substrates with compatible lattice constants such as GaN,[38, 39] mica[37], and sapphire.[35, 50]

Sapphire has been widely used as a substrate for the growth of 2D TMDs layers owing to its special lattice constant and flat insulating surfaces. Many reports have adopted c-plane sapphire as the substrate for growing TMDs layers.[17,30,31] However, it is known that the sapphire surface can be reconstructed at high temperatures to form long terraces and wide steps.[91] Chen, *et al.* have reported that the step-edges formed on the c-plane sapphire guided the aligned growth of WSe_2,[50] where the alignment became prominent only at the temperatures as high as 950°C. Meanwhile, Dumcenco, *et al.* reported that the key to achieving orientation alignment is the atomically smooth sapphire surfaces, where the sapphire needed to be annealed at a temperature of 1000°C in the air for 1 h just before the growth process.[35] However, several other reports did achieve the aligned TMDs triangles without sapphire pre-annealing at such a high temperature prior to the growth.[9] Apparently, the fundamental details of aligned growth are still not yet fully understood. Since many parameters in CVD also affect the growth of TMDs, such as precursor ratio, pressure, temperature, a systematic study to understand the mechanism of orientation alignment is required. In this study, we report that step-edge or terrace formation on sapphire is not the deterministic factor for the aligned growth of MoS_2. The highly oriented MoS_2 monolayers can be achieved at a much lower temperature (750 °C). The higher ratio of sulfur to MoO_3 at the initial nucleation step leads to a relatively smaller size of seeds spontaneously formed on sapphire. Consequently, these small seeds can easily rotate to the energetically favorable position, which is determined by the lattice structure of the substrate, leading to preferred orientation and alignment of TMDs. By contrast, a lower sulfur to MoO_3 ratio typically results in a larger size of seeds, which loses the capability of orientation control. High-resolution transmission electron microscopy (HRTEM) and energy dispersive X-ray analysis (EDX) reveal that the seeds are also crystalline MoS_2 layers. Density functional theory (DFT) simulations confirm that the small MoS_2 seeds exhibit preferred orientation on c-plane sapphire. This work provides a fundamental understanding of the seed formation and mechanism for orientation-controlled growth of TMDs monolayer.

3.2 Methods

3.2.1 Chemical Vapor Deposition of TMDs.

The MoS_2 monolayer was grown by the chemical vapor deposition method. The precursors are MoO_3 (Sigma-Aldrich, $\geq$ 99.5% purity) and S (Sigma-Aldrich, $\geq$99.5% purity) powders. The sulfur powder (4 g) was put at the upper stream side of the furnace (heating zone 1), and the temperature was maintained at 140 °C during the reaction. In the heating zone 2, center of the reaction chamber, the MoO_3 powder (0.8 g) was placed in a ceramic boat with a 1 cm × 5 cm sapphire substrate placed at the downstream of the ceramic boat. The gas flow was from Ar (Ar = 90 sccm) and the chamber pressure was controlled at 40 Torr. First, the center of the furnace was gradually heated from room temperature to 750 °C at a ramping rate of 25 °C/min and kept at this temperature for 5 min. Then, the temperature was increased to the growth temperature 800 °C with the same ramping rate and kept for 10 min. The furnace was then naturally cooled to room temperature. The consumption rate for S and MoO_3 is estimated as 1.69 mg/min and 0.88 mg/min respectively (Sulfur-rich reaction condition).

3.2.2 Characterization

Optical Images were collected using a Witec alpha 300 confocal Raman microscope with a RayShield coupler. The Mo concentration was analysed using scanning electron microscopy (SEM) imaging and energy dispersive X-ray (EDX) spectra which conducted using FEI Quanta 600 EDAX operating at 10 kV. The seeds and surface morphologies were examined on a commercial multifunction AFM instrument (Cypher ES model from Asylum Research Oxford Instruments) operating in contact mode. Olympus (OMCL-AC240TS) Al-coated silicon cantilevers were used for AFM characterizations. The resonance frequency was ~70 kHz; the spring constant was ~2 N/m, and the tip curvature radius was ~7 nm. The TEM cross-sectional samples were prepared in a Helios NanoLab 660 DualBeam FIB(focus ion Beam) system. Cross section HRTEM imaging and energy dispersive X-ray (EDX) spectrum mapping data were conducted using FEI TITAN AND OSIRIS operating at 300 kV and 200kV respectively.

3.2.3 Density Functional Theory Modeling

The calculations started from the optimization of an Al_2O_3 slab consisting of 4 by 4 unit cells with 5 layers of oxides. The optimization were performed using the RPBE functional[92] and the projected-augmented plane-wave method[93] using the VASP package.[94] The vdW interactions were described through Grimme's correction.[95] The energy cutoff was chosen as 400 eV. The energy convergence criteria was chosen as 1.0×10^{-6} eV, and the force convergence criteria was chosen as 1.0×10^{-2} eV/Å. Due to the computational load, the scanning process for the PES with different orientation angles uses 1x1x1 k point (at the □ point), while the energy calculation for the chosen configurations uses 4x4x2 k points using a Monkhorst-Pack grid. The scanning for Figure 5a is performed at a 0° fixed angle over a 7 × 7 Å² region with a 0.2 Å grid length to cover larger area (longer than the terminal oxygen-oxygen distance). The scanning for other angles is to find the nearby energy minima, therefore, only cover a smaller 3×3 Å^2 area with the 0.1 Å grid length.

3.3 Results and Discussion

3.3.1 Effect of S/MoO$_3$ precursor ratio on MoS$_2$ alignment

We adopt the CVD growth process first reported by Lee, *et al.* where the growth of MoS$_2$ monolayer on sapphire relies on the gas-phase reaction of MoO$_3$ and sulfur vapors carried by a pure Ar flow.[8] Sapphire is commonly used for MoS$_2$ growth since it is an insulating substrate with high thermal stability and excellent crystalline quality. Additionally, sapphire and MoS$_2$ both share a hexagonal crystal structure, which makes the growth of TMDs on sapphire preferable. For this reason, the structure of the sapphire surface plays a key role in determining the MoS$_2$ orientation through van der Waals (vdW) interaction.[35] It is well known that various surface treatments on sapphire lead to different termination layers on its surfaces, *e.g.,* Al, OH, or O termination,[96] which may strongly affect the results of the MoS$_2$ growth. For our MoS$_2$ growth, sapphire is first treated with a piranha solution, resulting in OH group termination on the surfaces. Upon heating during the growth, two neighboring Al-OHs shall be dehydrated into Al$_2$O$_3$[97] and thus the MoS$_2$ is grown on the sapphire with an oxygen-terminated surface.

Figure 3.1a schematically illustrates the growth of MoS$_2$ layers on c-plane sapphire substrates using the CVD process, and the heating profiles of the precursors sulfur and MoO$_3$ powders are shown in Figure 1b. Other growth details are provided in the method section. In our typical CVD process, the sulfur powders (at heating zone 1) were first heated to 140°C to fill up the reaction tube with S vapors, followed by the temperature ramping of MoO_3 powders and sapphire substrates (at heating zone 2) to the growth temperature of 800 °C as illustrated in the heating profile T_{2A} in Figure 3.1b. Interestingly, we observe drastically different growth behaviors at different locations of the same sapphire substrate. Figure 3.1c presents a schematic illustration of how the growth substrate is spatially separated into three locations. Figure 3.1d shows the relative concentration of MoO_3 at these locations based on the analysis of our separate experiment which determines the relative concentration of Mo deposits on substrates using EDX (details in Appendix A, Table A1, and Figure D1). The results clearly show that MoO_3 vapor concentration fast decreases with the distance away from the MoO_3 source. Note that in our experiments the sulfur vapor concentration is in excess relative to MoO_3 inside the tube. Hence, our experiments were different from those described by Govind Rajan, *et al.*,[98] where MoO_3 concentration was constant but S concentration decays along the reaction tube. In addition, the diffusivity of sulfur vapors is much higher than that of MoO_3.[99] As a result, the S/MoO_3 ratio significantly increases with distance from the MoO_3 source. The corresponding optical microscope (OM) images for the MoS_2 monolayer flakes growing on these locations are shown in Figure 3.1e. In the location I, the MoS_2 flakes are randomly oriented but they start to show some alignment at location II. In location III, these MoS_2 flakes exhibit dominant edge orientation (0° and 60° as detailed in the statistical analysis in Appendix A Figure A1). The effect is not caused by the temperature variations since the temperatures at these locations are the same (within the range of ± 1 ℃). In previous reports, incrementing the S/MoO_3 ratio results in the geometry of the produced MoS_2 shifting from hexagon to triangle.[85, 86, 98] Our experiments always produce triangular MoS_2 flakes, indicating the excess amount of sulfur vapor in the reaction tube. Besides, our experimental results clearly correlate the S/MoO_3 vapor ratio to the orientation control. We hypothesize that at a high S/MoO_3 ratio the sulfur vapors efficiently reduce the MoO_3 to form small MoS_2 crystalline seeds, which have the ability to rotate and align with the lattice structure of the substrate. By contrast, a low S/MoO_3 ratio may result in the incomplete sulfurization of MoO_3, with relatively larger and thicker seeds. These sub-oxide nanoparticles may land on the sapphire substrate randomly. Consequently, the synthesized MoS_2 flakes exhibit random orientations.

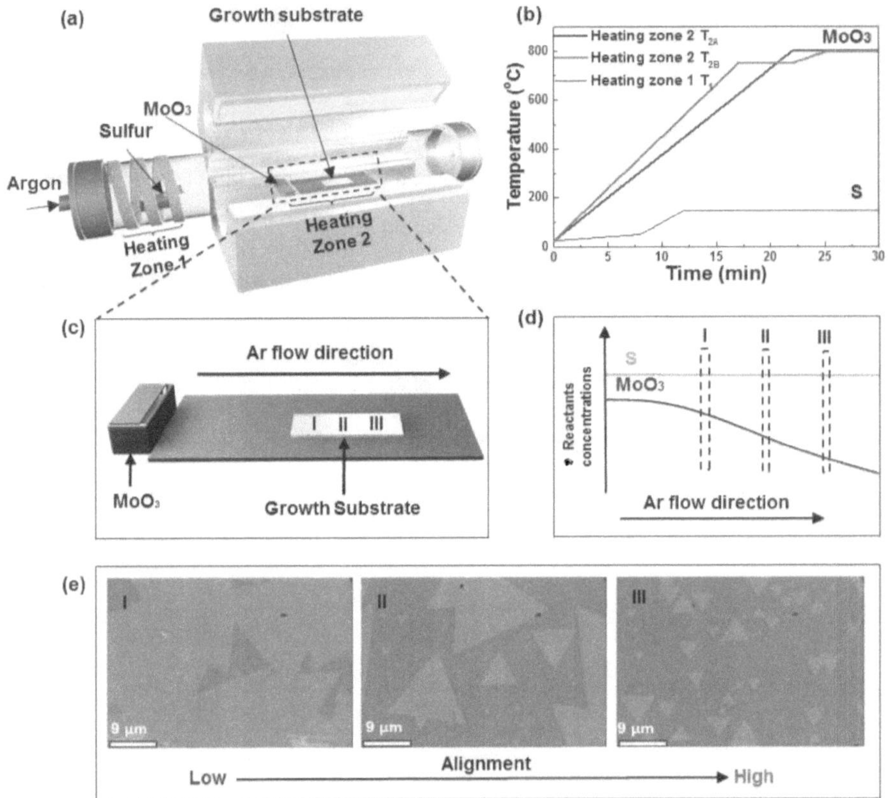

Figure 3.1 Effect of S/MoO3 precursor ratio on MoS2 alignment. (a) Schematic illustration of the experimental setup of the MoS$_2$ growth. (b) The temperature heating profiles were adopted for the study of growing MoS$_2$. (c) Schematic illustration for the various locations (I, II, and III) according to the distance away from the MoO$_3$ source. (d) A schematic illustration of the concentration of the reactants of MoO$_3$ and S reached the specified locations in (c). (e)The corresponding OM images of the MoS$_2$ flakes at each location. Scale bars: 9 µm.

3.3.2 Growth of aligned MoS$_2$ monolayers with a seeding step

Since a sulfur-rich environment is essential to form small MoS$_2$ seeds as hypothesized above, the two-step heating profile T$_{2B}$ (shown in Figure 3.1b) is designed to achieve better alignment across the whole sample regions (including locations I, II, and III), where the substrate first stay at 750 °C for 5 min before it is heated to the growth temperature. This step ensures a highly S-rich environment for completing the sulfurization of slightly evaporated MoO$_3$ at the seed formation stage. Consequently, very small seeds can form and rotate easily to stay at a more energetically favorable orientation on the substrate. Although growing MoS$_2$ flakes can be very slow at 750 °C due to insufficient MoO$_3$ vapors, extending the growth time results in small monolayers with edge lengths less than 1 μm as shown in Appendix A Figure A3. This observation is consistent with Pan *et al.*, where they revealed that at the low temperature of 750 °C tiny monolayers on Si substrate were obtained.[100] Therefore, we designed a profile where after aligning small seeds at 750 °C, the temperature is increased to 800 °C to increase the lateral growth and enlarge the monolayer's size. Gratifyingly, we observed that all regions including I, II, and III are grown with aligned MoS$_2$ monolayers, see Figure 3.2. Figure 3.2a shows the typical OM images of the MoS$_2$ monolayer flakes grown at the central region II using the heating profiles T$_{2B}$, where the MoS$_2$ flakes are highly aligned and the dominant edge orientations are 0° and 60° as illustrated by the statistical analysis in Figure 3.2b. For comparison, one selected typical OM image for the MoS$_2$ monolayers growth with the typical T$_{2A}$ profile (region II) is shown in Figure 3.2c and the corresponding statistical orientation analysis in Figure 3.2d clearly demonstrates the feature of random orientation. We have also performed separate experiments to examine the importance of the sulfur-rich condition on MoS_2 alignment by changing the feeding time of sulfur. The results in Appendix A Figure A4 show that early feeding of S vapors results in better orientation alignment, strongly corroborating our hypothesis.

At a high temperature, c-plane sapphire would usually develop terrace structures with atomic steps on the surface. Aligned growth as guided by those steps has been reported for WSe_2 on sapphire, where the growth temperature was above 950 °C.[50] However, in our case the growth temperature of 800°C is not sufficiently high to conduct step edge-guided aligned growth. Consequently, our growth orientation is controlled by the lattice crystal structure of the substrates instead of the terraces. Figure 3.2e shows an atomic force microscope (AFM) image for a monolayer MoS_2 grown by our process on a sapphire substrate, where the edge of the flake is clearly not aligned with the steps from terrace structures.

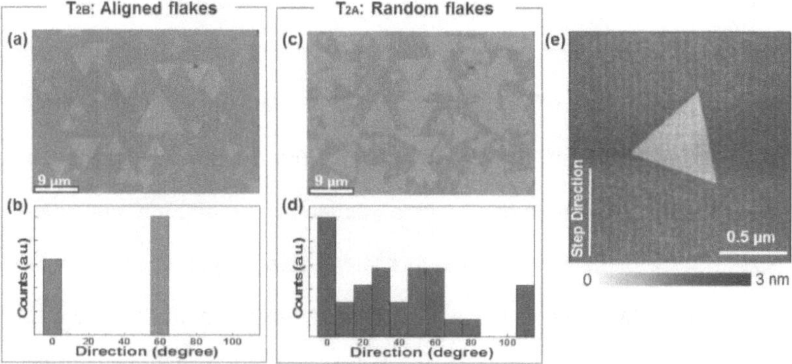

Figure 3.2 As-grown MoS₂ flakes using different heating profiles. (a) The OM image of the as-grown MoS_2 flakes using a T_{2B} heating profile. (b) Histograms of the orientation distributions based on the image (a). (c) The OM image of as-grown MoS_2 flakes using the hT_{2A} heating profile. (d) Histograms of the orientation distributions based on the image (c). (e) AFM image of MoS_2 monolayer flakes grown on c-plane sapphire using a T_{2B} heating profile.

3.3.3 Structure of the seeds

To explore the structure of the seeds, the growth is stopped after the seed formation stage (750 °C for 5 min in a sulfur-rich environment). AFM is adopted to characterize the morphology and size of the seeds initiated at this stage. Figure 3.3a is a typical AFM image for the seeds formed at the upstream side of the reaction zone. The shape of the seeds is identified as a triangle (more AFM images of the MoS_2 seeds are provided in Appendix A Figure A5). Unlike the results for the one-step heating profile, the triangles in the upstream are aligned since the two-steps heating profile results in small seeds in all the substrate locations which can rotate and align to the substrate. Figure 3.3b shows the cross-sectional height profile for the seed along the dashed line in Figure 3.3a, given a thickness of ~0.6 nm monolayer which agrees well with the reported thickness of MoS_2 monolayers.[101] Figure 3.3c displays the AFM image for the seeds formed at the downstream side with a thickness of ~1.2 nm (Figure 3d). The seeds at the downstream side are normally with an apparent size of 20-30 nm. Note that the size may be overestimated since the lateral size measurement is limited by the size of the AFM tip end. Hence, identifying the shape of the seed is challenging. The size difference of the seeds in different regions may be related to the amount of the MoO_3 vapor reaching the regions. It would require future investigations.

Most of the as-grown MoS_2 monolayers do not exhibit an identifiable nucleation center in the middle of the triangles under OM observation. However, we still occasionally find thick seeds with a pyramid shape as shown in Figure 3.4a. The cross-sectional HRTEM image in Figure 3.4b confirms these thick seeds consist of stacked MoS_2 layers, indicating even larger MoO_3 can be completely sulfurized to form MoS_2. The thickness is measured to be 7.5 nm corresponding to around 10 layers of MoS_2. Additionally, Figure 3.4c shows the energy dispersive X-ray analysis (EDX) images for Al, O, Mo, and S signals at the same location in Figure 3.4b. No O is observed in the MoS_2 multilayers, which again confirms that the seed is composed of stacked MoS_2 and the precursor MoO_3 is completely sulfurized at the seed formation stage.

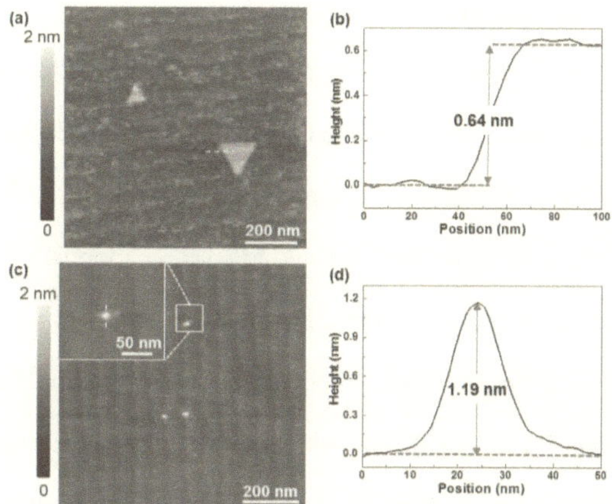

Figure 3.3 MoS₂ seeds. (a) AFM image of the MoS2 seeds in the upstream region. (b) Height profile for the white dashed line in (a). (c) AFM image of the MoS2 seeds in the downstream region. The inset in (c) shows a zoomed AFM image in the white box. (d) Height profile for the white dashed line (c).

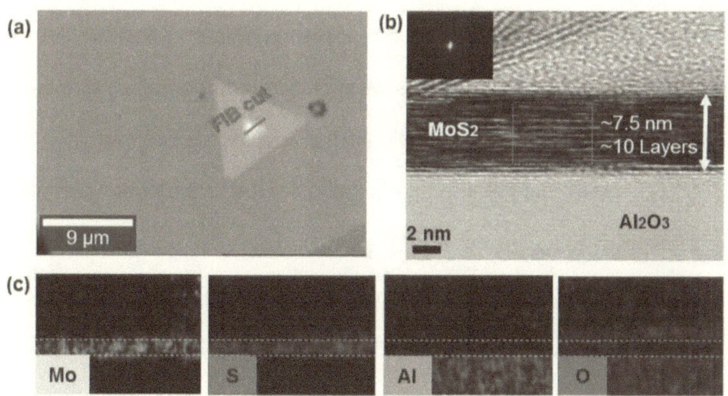

Figure 3.4 Cross-sectional HRTEM of MoS₂ on c-plane sapphire demonstrating the seed composition. (a) The OM image of as-grown MoS₂ flakes with a thick seed. The red line marks the area from which the FIB was cut; (b) Cross-sectional HRTEM image of MoS₂ and (c) the corresponding EDX maps of (b) for Mo, S, Al, and O signals.

3.3.4 Density functional theory simulation

A more detailed investigation from the atomic viewpoint was performed through density functional theory (DFT) calculations. A model consisting of 4×4 unit cells of Al_2O_3 and 6×6 MoS_2 was constructed to minimize the lattice constant mismatch to 0.7%. As described, we believe the MoS_2 monolayer is deposited on the sapphire c-plane with oxygen terminations, which is in clear contrast to the previously proposed models using the sapphire terminated with H or Al.[35] After optimization (Figure 3.5a), the vertical distance between the upper sulfur of MoS_2 and the sapphire surface is 5.8 Å, which is comparable with our AFM characterization.

Since the seed only contains Mo and S, it is represented by a more prototypical hexagon MoS_2 piece as shown in Figure 3.5b. A systematic scanning consisting of over 7000 calculations with fixed relative orientations and positions between the MoS_2 seed and Al_2O_3 substrate were performed, thereafter, to evaluate potential energy surface (PES) as a function of the relative position and orientation between the MoS_2 seed and the sapphire substrate. Due to the heavy computational load, the distance between the top O layer of Al_2O_3 and the top S layer of MoS_2 was fixed at 5.8 Å, which is the distance obtained from structural optimization. Figure 3.5c gives a description of the PES corresponding to an angle of 0°, and the other maps are shown in Appendix A Figure A6. We have scanned large enough area to cover the Al_2O_3 surface region, and the distance between the maxima and minima of the PES is consistent with the distance between oxygen atoms at the c-plane, indicating that the vdW interaction can tune the relative position of the MoS_2

seeds landing on the surface. Thereafter, the configurations corresponding to the minimum in this PES is chosen for orientation scans of the MoS_2 seed. For each specific orientation, the relative positions were also scanned to get a complete view of the PES. The representative configurations for each orientation were chosen from the minimum of the PES, and the energies were calculated with higher precision (details in Appendix A). We obtain relative energy as a function of orientation angles with the most favorable position of the MoS_2 seed as shown in Figure 3.5d. Considering the symmetry feature of our model, the results corresponding to 0~60 degree relative orientation were demonstrated in this report.

Even though we simplify the simulation by using this small hexagon seed without considering the structural variation induced by high experimental temperature, the simulation can give a consistent trend compared with experiments: the 0° or 60° configurations are most favorable. It is also of interest to obtain further insights from this model. For instance, the PES is quite smooth and does not have large kinetic barriers (~5 kBT for experimental temperature), indicating the seed can smoothly find its energetic minimum. Consequently, in our experiment, a 5-minute plateau at 750 °C (Figure 3.1b) plays a role in assisting the small seeds in finding their favorable configurations. Besides, any treatment destroying this smooth PES by generating additional large energy barriers may bring troubles by kinetically trapping the seed into a local minimum. In short, a relatively flat substrate surface having periodic energetic minimums without significant energy barriers can be suitable to template the seed and generate highly aligned TMDs monolayers.

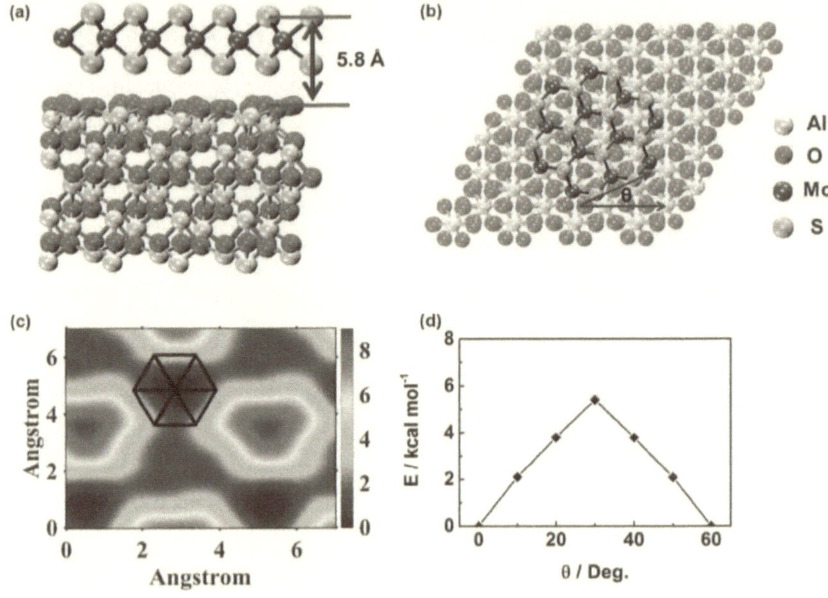

Figure 3.5 DFT calculation of the relative energy as a function between different orientation angles between MoS₂ seed and sapphire substrate. (a) The optimized MoS₂ layer on the sapphire substrate demonstrates. (b) Schematic view of the MoS₂ hexagon seed on the sapphire substrate. (c) Potential energy surface as a function of the different relative positions of MoS₂ seed on the sapphire substrate when the relative angle is fixed as 0. (d) Potential energy as a function of the relative angle between the MoS₂ seed on sapphire; each configuration was chosen as the minimum point on the position scan (0 degree in c, the rest in supporting information).

3.4 Conclusion

In conclusion, we have synthesized highly oriented MoS_2 monolayers by controlling the sulfurization of MoO_3 using the CVD method. It is concluded that high sulfurization in the seed formation stage plays an important role to form a small seed, which can easily align with the lattice structure of the sapphire substrate. Seeds can be grown at 750 °C, whereas a higher temperature of 800 °C speeds up the lateral growth. Additionally, we provide evidence that the seed is completely sulfurized and consists of MoS_2. Such orientation control can be applicable for other 2D TMDs growth on a crystalline substrate. It is anticipated that such understanding of growth mechanism is crucially important for controlling the growth of diverse 2D materials and further understanding the aligning mechanism of TMDs growth on other hexagonal surfaces such as TMDs, BN, graphene, and graphite heterostructure or single crystal substrates with compatible lattice constants such as GaN and mica.

Chapter 4: Ledge-Directed Epitaxy of Continuously Self-aligned Single-crystalline Nanoribbons of Transition Metal Dichalcogenides

4.1 Introduction

Planar transistors have been used for myriad generations with size and voltage scaling to enhance performance and save cost, following the well-known Moore's Law[102]. The innovation of Fin-field effect transistor (Fin-FET) architecture was the solution and rendered the further device scaling possible. Unfortunately, the short channel effect shall ultimately limit the Fin-FET scaling. A wave of revolutionary design in FET architecture with a superior gate control over the channel begins to take hold. This emerging stacked sheet architecture typically consists of multi-stacked semiconducting nanosheets with surrounding gate metals, which demonstrates better short-channel control and thus holds promise to extend the Moore's Law[103]. Aligned arrays of single-crystal, monolayer 2D TMDs nanoribbons with high aspect-ratios, which represent the ultimate limit of miniaturization in the vertical dimension, are therefore very attractive in this context. Specifically, the ability to achieve single crystallinity and electrical uniformity throughout the entirety of the 2D TMDs nanoribbons, the key metrics of enabling batch production FET arrays, would allow a very high degree of electrostatic control at very low power consumption. Synthetic strategies, toward TMDs nanoribbons, have been reported to individually achieve control of layer number, single-crystallinity, self-alignment, and dimensionalities[27, 48, 104]. However, the dearth of a manufacturing route toward

TMDs nanoribbons that synergistically combine all the aforementioned properties remains a major challenge.

4.1.1 Transfer of Monolayer MoS$_2$ nanoribbons

After the CVD growth, the resulting MoS$_2$ nanoribbons on β-Ga$_2$O$_3$ (100) were transferred onto a substrate of interest via a polydimethylsiloxane (PDMS)-assisted approach. In brief, a thin PDMS film was placed on top of MoS$_2$/β-Ga$_2$O$_3$. Note that it is very critical to ensure a conformal contact between PDMS and MoS$_2$/β-Ga$_2$O$_3$. Next, the PDMS/MoS$_2$/β-Ga$_2$O$_3$ stacked film was soaked in a 1M KOH for 5 min at room temperature, followed by rinsing the sample with copious amount of deionized (DI) water, PDMS/MoS$_2$ stacked film was slowly peeled-off from β-Ga$_2$O$_3$ and then placed on a target substrate. The sample was kept in a vacuum for 30 min to make sure the adhesion between MoS$_2$ and the target substrate. Residual water droplets were dried under a constant nitrogen (N$_2$) flow. Finally, PDMS was peeled off, leaving behind the MoS$_2$ nanoribbons on the target substrate.

4.1.2 Characterizations

A FEI Quanta 600 scanning electron microscopy (SEM) was utilized to provide morphological views operating at 5 kV. Raman and photoluminescence (PL) spectra on MoS$_2$, WSe$_2$ nanoribbons and the lateral heterostructures of both were collected using a Witec alpha 300 confocal Raman microscope equipped with a RayShield coupler. A 532-nm solid state laser as the excitation source. The excitation light with a power of 2.5 mW was focused onto the sample by a 100X objective lens (N.A. = 0.9). The signal was collected by the same objective lens, analyzed by a 0.75-m monochromator and detected by a liquid-nitrogen-cooled CCD camera. The atomic force microscopy (AFM) characterizations was conducted with Olympus (OMCLAC240TS) Al-coated silicon cantilevers. The resonance frequency was ~70 kHz, the spring constant was ~2 N/m, and the tip curvature radius was ~7 nm. The C-AFM measurements were conducted using Pt-Ir-coated conductive probes (SCM-PIT, Bruker) with a spring constant in the range of 0.5 to 4Nm^{-1}. Contact mode was utilized for imaging. A constant tip to sample bias of +3V was applied during all the C-AFM scans. AFM raw images were processed using a Gwydion 2.51 software. The cross-sectional scanning transmission electron microscopy (STEM) samples were prepared in a Helios NanoLab 660 DualBeam focused ion beam (FIB) system. Cross-sectional HR-STEM imaging was conducted using a Thermofisher USA (former FEI) Titan Themis Z transmission electron microscope (TEM) equipped with a double Cs (spherical aberration) corrector operating at 300 kV. Dark field (DF)-STEM with High-angle annular dark-field (HAADF) imaging was done in a scanning mode using an acceleration voltage of 80 kV with a column (at the sample) vacuum of about 2–4X10^{-7} Torr at room temperate. Hyper-spectral photoluminescence (PL) measurements and mapping were taken with an IMA™ hyperspectral microscope from Photon Inc. The samples were excited from a 532 nm laser with the intensity of 6.4 µW/µm^2 and spectra were collected from an area of 90 µm x 65 µm with 1 nm resolution. Exposure time was chosen as 120 sec for each wavelength.

4.1.3 Second Harmonic Generation (SHG) measurements

A home-built microscope arranged in the backscattering set-up was utilized to measure the SHG signals. Laser pulses generated from a mode-locked titanium (Ti)-sapphire laser with a wavelength of 850 nm, a pulse width of ~150 fs and a repetition rate of 80 MHz were used as the fundamental laser field. The polarization was selected by a linear polarizer and a half waveplate. The laser beam which illuminates normally onto the sample was focused by a 100X objective lens with a N.A. of 0.9. The generated SHG signals were collected by the same objective and sent to a 0.75 m spectrometer equipped with a liquid N_2 cooled CCD camera. The scattered fundamental laser field was blocked by a 450 nm short pass filter. For the measurement of polarization resolved SHG images, the sample was mounted on a motorized x-y stage and scanned with a step size of 0.2 μm. The SHG signals were sent through a polarizing beam-splitter to separate the SHG intensity with polarizations that are parallel and perpendicular to the laser polarization. The two components were sent into the spectrometer and detected by the CCD camera simultaneously. For the polarization resolved SHG, the orientation of MoS_2 nanoribbons was aligned closely to the polarization direction of the excitation laser. The linear polarization of the excitation laser and the SHG signals were selected and analyzed by the combination of a half waveplate and a linear polarizer. The polarization of the generated signals was rotated by a half waveplate mounted on a motorized rotational stage with a step of 4° and analyzed by the polarizer.

4.1.4 First-principles Calculation

The first-principles calculations were carried out based on the density function theory (DFT) as implemented in the Vienna Ab-initio Simulation Package (VASP)[105] within the MedeA[106]. The texchange-correlation potential described by the PBE-GGA[107], and the van der Walls (vdW) correction vdW-DF (optB86b) functional[108] is used to calculate the binding energy difference between MoS_2 molecules and β-Ga_2O_3 substrates. The β-Ga_2O_3 has a monoclinic structure with lattice constants of a = 3.037 Å, b = 5.798 Å, and β = 103.8°[109]. A k-grid of 8×1×1 and an energy cut-off 400 eV were used for the system of $5MoS_2/1×5$ β-Ga_2O_3, with a vacuum of 20 Å along c and at least 10 Å along b between MoS_2 molecule and the other side of ledge in periodic cell to eliminate spurious interaction.

4.1.5 Field-Effect Transistor Fabrication and Measurement

 Monolayer MoS_2 nanoribbons grown on β-Ga_2O_3 (100) were transferred on 15 nm HfO_2, which was deposited on heavily doped silicon (Si) via the atomic layer deposition (ALD) as a gate insulator. A single crystalline hBN monolayer was detached from the Cu (111)/sapphire substrate by electrochemical delamination and then transferred onto HfO_2/Si via a combination of thermal release tape (TRT; #3195M) and poly (methyl methacrylate) (PMMA). The TRT can be released by annealing the TRT/PMMA/hBN/HfO_2/Si stacked films on a hotplate at 180°C. The PMMA film was thoroughly removed via iteratively immersing the sample in a hot acetone bath for 40 min, leaving behind a hBN/HfO_2/Si stacked substrate. After transferring the MoS_2 nanoribbons, the resulting MoS_2 nanoribbons/hBN/HfO_2/Si were placed in a vacuum chamber under a pressure of 10^{-6} torr for 12h. Owing to the globe alignment of LDE grown MoS_2 nanoribbons that provides far fewer constraints for the effective fabrication of field effect transistors, electron-beam lithography emerges as the reliable method for producing the patterns of metal electrodes comprised of nickel (Ni, 20 nm) and gold (Au, 50 nm) necessary for electrical testing. More than one hundred single-nanoribbon field-effect transistors were reliably produced, and all tested to confirm the electrical output performance. This is due to the uniform, self-aligned, and tunable distribution of MoS_2 nanoribbons over the entire area of the β-Ga_2O_3 (100) substrate (~1 cm x 1.5 cm). All measurements were carried out under ambient conditions using a Keithley 4200 semiconductor analyzer.

4.2 Results and Discussion

4.2.1 LDE growth of wafer-scale, globally aligned monolayer MoS_2 nanoribbons

It is known that the lattice orientation of 2D TMDs can be guided by substrates through lifting the energy degeneracy of the 2D TMD-substrate van der Waals (vdW) system[44]. We have further revealed that the lateral docking of 2D hexagonal boron nitride (hBN) seeds to the atomic step edges of Cu (111) substrates predominate over the vertical vdW registry of hBN on Cu, ensuring the mono-orientated nucleation and thus achieving the growth of single-crystal 2D hBN film[52]. These neat demonstrations of synthesizing the uniform monolayer 2D TMDs films with single crystallinity highlight that the selection of substrate (e.g., thermodynamics) and the growth parameter control (e.g., kinetics) are critically important. Here, we explore the epitaxial growth of single-crystalline and aligned TMDs nanoribbons via LDE assisted chemical vapor deposition (CVD), that hinges on the thermodynamic control of TMDs seeding orientation in conjunction with the kinetic control of growth direction. Figure 4.1a illustrates the LDE growth scheme for MoS_2 nanoribbons: (I) A single-crystal β-Ga_2O_3 (100) substrate with exposed ledges is used; (II) Nucleation of MoS_2 seeds with preferred orientation takes place at the ledges of β-Ga_2O_3; (III) Aligned MoS_2 domains merge into continuous nanoribbons; (IV) MoS_2 nanoribbons can be easily peeled off from the β-Ga_2O_3 (100) substrate and readily transferred to arbitrary substrates via a polydimethylsiloxane (PDMS)-assisted process; and (V) Exfoliated β-Ga_2O_3 substrate can be re-used for another round of growth.

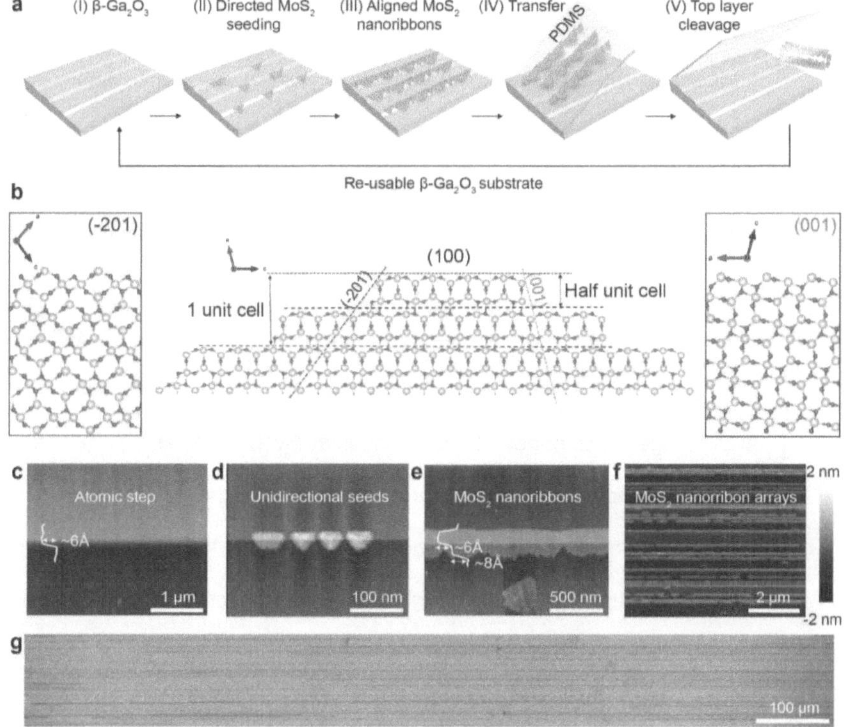

Figure 4.1 LDE growth of wafer-scale, globally aligned monolayer MoS₂ nanoribbons. (a) A schematic illustration depicts the sequential growth of the monolayer MoS₂ nanoribbons along the intrinsically aligned ledges on the β-Ga₂O₃ (100) substrate, which can be reused after a facile mechanical exfoliation. (b) Computer-generated crystal structures provide a cross-sectional view of both (-201) and (001) ledges on the (100) β-Ga₂O₃. (c) Height profile along the well-defined atomic step on pristine β-Ga₂O₃ substrate helps determine the step height of ~6 Å. (d) Nucleation of unidirectional MoS₂ seeds along the ledge. (e) A continuous MoS₂ nanoribbon with asymmetric edges. (f) Dense arrays of aligned MoS₂ nanoribbons. (g) SEM image showcases the continuous growth of globally aligned MoS₂ nanoribbons beyond the millimeter-scale.

4.2.2 Polarization-resolved SHG measurement and DF-STEM characterization

Intrinsically, the (100) plane of the freshly exfoliated β-Ga$_2$O$_3$ substrate exhibits atomically sharp steps with a step height of ~6 Å (half unit cell). These steps trend up and down across the entire β-Ga$_2$O$_3$ substrate, resulting in two sets of structurally equivalent but crystallographically inverted ledges namely (001) and (-201)[110, 111] (Figure 4.1b). As featured in Figure 4.1c-e, various stages in the growth of MoS$_2$ nanoribbons are revealed by atomic force microscopy (AFM). Figure 4.1c clearly suggests that the unidirectional nucleation of four MoS$_2$ seeds carries out at the ledges. The edges of these triangular MoS$_2$ seeds stay parallel to the well-defined step edge whereas the vertices point toward the lower terrace. Meanwhile, we observed that the nucleation density of oriented MoS$_2$ seeds along both (001) and (-201) ledges is overwhelmingly higher than that of flat terraces where the only sporadic distribution of randomly oriented MoS$_2$ flakes (orientation varies between 0°, 90°, 180° and 270° owing to the symmetry of β-Ga$_2$O$_3$ substrate, which is monoclinic in nature) can be spotted. The observation of unidirectional MoS$_2$ seeding flakes on the atomically-textured, single-crystalline β-Ga$_2$O$_3$ (100) substrate clearly indicates the existence of an energetically minimized MoS$_2$—β-Ga$_2$O$_3$ ledge configuration, thus forming the basis for subsequent coalesce into continuous nanoribbons with single crystallinity. Indeed, aligned and mono-oriented MoS$_2$ seeds grow by successive addition from the surrounding precursors and ultimately merge into a MoS$_2$ nanoribbon as LDE approaches completion (Figure 4.1e). The resulting MoS$_2$ nanoribbons exhibit a uniform step height of ~8 Å, characteristic of monolayer MoS$_2$. Another unique capability of LDE is the controlled nucleation and unidirectional growth of ordered arrays of MoS$_2$ nanoribbons at the atomic scale in previously unachievable quantities and scales (up to a centimeter long and aspect ratios > 5,000). Images of AFM and scanning electron microscopy (SEM) collectively demonstrated the growth of dense arrays of globally aligned, continuous MoS$_2$ nanoribbons enabled by LDE over the entire β-Ga$_2$O$_3$ (100) substrate (Figure 4.1f-g).

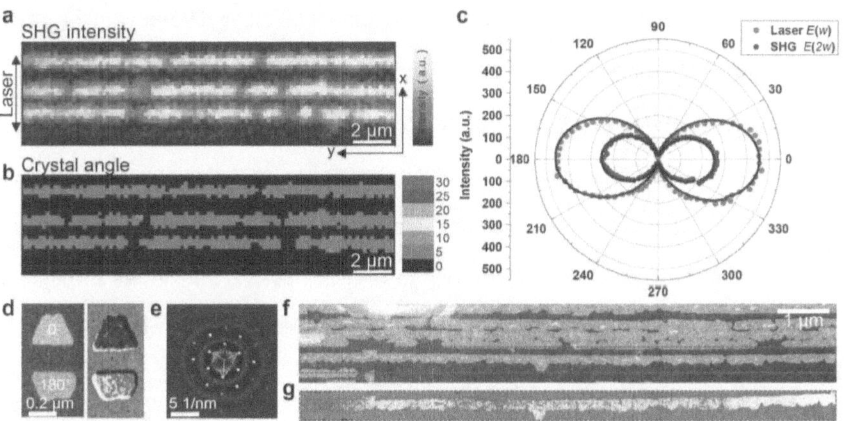

Figure 4.2 Polarization-resolved SHG measurement and DF-STEM characterization. (a) The intensity map of the parallel component of SHG. The black arrow indicates the direction of incident laser polarization. (b) Map of the angle (θ) between the direction of laser polarization and the armchair direction of the MoS$_2$ nanoribbons. (c) Polar plot of the polarization-resolved SHG intensity and the backscattered laser light as a function of detection angles. (d) (left) MoS$_2$ mirror domains of 0° and 180° orientations can be clearly juxtaposed with the differential diffraction filtered STEM mapping of domain orientations (right). (e) Corresponding convergent beam diffraction patterns of the oppositely oriented domains in (d). (f) ADF-STEM image together with (g) differential diffraction filtered STEM mapping of the MoS$_2$ nanoribbon highlighted in the red dotted rectangle collectively attests to the absence of grain boundaries which in turn confirms the single crystallinity of the LDE MoS$_2$ nanoribbons.

In parallel, the innate step edges which are naturally present on the monolithic β-Ga_2O_3 (100) crystals[112], have a propensity to cleave parallel to (100) and (001) planes by half unit cell. This is the result of unique octahedral arrangements of the Ga atoms (see Figure B1) which are parallel to the (010) plane[109]. Consequently, the newly exfoliated (100) plane of β-Ga_2O_3 retains atomically clean, ordered and spatially distributed step edges with half-unit cell ledges as shown in Figure 4.1b[110]. Photoluminescence (PL) measurement taken on different batches of MoS_2 nanoribbons grown on the repeatedly exfoliated β-Ga_2O_3 (100) substrate reveals neither changes in full width at half maximum (FWHM) nor the shift in PL peaks (see Appendix B Figure B2), making it possible for the continuous and reliable batch production of high-quality MoS_2 nanoribbons. This peeling feature is particularly appealing as the ability to re-use β-Ga_2O_3 substrates eliminates the needs for a time-consuming and often laborious lithography process.

While all the aligned MoS_2 flakes interlock in the same way and have the identical orientation, the atomic structures of β-Ga_2O_3 (100) has a profound implication on the geometric shapes of edges of MoS_2 nanoribbons. Unlike the MoS_2 flakes grown on symmetrical substrate, MoS_2 flakes grown on β-Ga_2O_3 (100) exhibit asymmetrically shaped edges, i.e., smooth and zigzag-shaped edges. Away from the well-defined ledges, the extremities of the merged MoS_2 flakes are permitted to grow without any external constraint. The one edge of the single crystalline MoS_2 nanoribbons assumes a regular zig-zag shape. High-angle annular dark-field scanning transmission electron microscopy (HAADF-STEM) Images in Appendix B Figure B3 confirm the zigzag-shaped edges of the nanoribbons. Occasionally, we observed the formation of bilayer MoS_2 nanoribbons. High-resolution (HR) HAADF-STEM images near the edges of the bilayer regions reveal the absence of Moiré patterns, indicating predominantly 2H stacking orders. Moreover, by controlling the growth temperature and nucleation density, the width of the MoS_2 nanoribbons can be systematically varied between 70 nm to 600 nm (see Appendix B Figure B4), for which the width likely can meet the requirement for stacked sheet transistor applications. Further decrease in width for which fundamental confinement effects may arise, such as changes in bandgap and presence of 1D metallicity (Appendix B Supplementary Discussion 1, and Figure B5 and Figure B6), is possible experimentally. Yet, the LDE TMDs nanoribbons may not be continuous in the millimeter to centimeter length scale.

To verify the orientation of individual flakes and the associated crystallinity of the MoS_2 nanoribbons, the LDE MoS_2 nanoribbons were characterized by the second harmonic generation (SHG) micro-spectroscopy and dark-field (DF) STEM. It is known that polarization-resolved SHG is sensitive to the crystal orientation and the intensity profile map can be used as a descriptor for verifying spatial orientations of the merged flakes within the coalesced nanoribbons[113-115]. Figure 4.2a displays the SHG intensity map taken from three horizontally aligned MoS_2 nanoribbons with perpendicular polarization. All three MoS_2 nanoribbons demonstrate homogenous SHG intensities except for a few nodes along the direction of laser irradiation. The discontinuity of SHG intensity is the result of rarely observed multilayer MoS_2 seeds interspersed between the continuous MoS_2 nanoribbons by comparison of AFM images (see Appendix B Figure B3 and Figure B7). The homogeneity of the SHG intensity proves that each nanoribbon indeed comprises MoS$_2$ flakes with a single orientation. Furthermore, we deduced the angles between the laser polarization direction and the nearest armchair direction via the equation: $\theta = (1/3)tan^{-1}\sqrt{I_x/I_y}$[113-115]. In this light, Figure 4.2b features the intensity map that spatially resolves the angle distribution derived from compiling the simultaneously detected I_x and I_y SHG intensity, revealing uniform yet narrow angular distribution of ~2°. The orientation of the zigzag direction is further confirmed by drawing comparisons of polarization-resolved SHG intensity between the MoS$_2$ nanoribbons and the reflected laser from the substrate. As indicated in the polar plot of Figure 4.2c, MoS$_2$ flakes with mirror domains of 0° and 180° orientations are angularly equivalent in terms of SHG intensity. The SHG can help to characterize the nanoribbons in a large area but further distinguishing such mirror domains requires other methodologies.

It is known that the variation in crystallographic orientations disturbs the structural continuity, i.e., the formation of grain boundaries. This disruption manifests signs of polycrystalline in annular dark-field (ADF)-STEM on the nanometer length scale[19] as shown in Figure 4.2d (see Appendix B Supplementary discussion 2). Mirror domains of 0° (color in blue) and 180° (color in yellowish gold) can therefore be clearly determined on the basis of convergent beam electron diffraction patterns (Figure 4.2e). ADF-STEM images shown in Figure 4.2f, g confirm the absence of mirror domains and thus the existence of crystallographic continuity of our LDE MoS$_2$ nanoribbons on the micrometer length scale. This nanoribbon consists of more than twenty mono-oriented flakes and all ADF-STEM images exhibit crystallographically coherent domains with no visible grain boundaries, confirming the single-crystal nature of LDE MoS_2 nanoribbons. Other characterizations, including SEM image, corresponding PL mapping of the characteristic excitonic direct gap emission of monolayer MoS_2, and signatures from Raman spectroscopy, again prove the structure continuity and crystallographic coherence of chemical states of MoS_2 nanoribbons (Appendix B Figure B8).

4.2.3 Mechanism of LDE growth of MoS_2 nanoribbons on the ledge of β-Ga_2O_3 (100)

To understand the preferred nucleation at the ledge and the controlled growth along the step of β-Ga_2O_3, we conduct the cross-sectional HAADF HR-STEM (Figure 4.3a, b) to provide the atomically resolved structures of both MoS_2 nanoribbon and the underlying β-Ga_2O_3 (100) substrate. Focus ion beam (FIB) was performed in the transverse direction of the MoS_2 nanoribbon (perpendicular to the [010] of β-Ga_2O_3). The atomic structures of MoS_2 nanoribbons are divided into three regions based on the locations: namely, (I) bottom terrace (left), (II) ledge (center), and (III) top terrace (right), respectively, allowing us to elucidate the relationship between epilayer and growth substrate. In agreement with the AFM image (Figure 4.1d), region (II), the center segment of the nanoribbons where the nucleation of aligned, triangular seeds takes place, is found to lay above the (-201) ledge of β-Ga_2O_3 (Figure 4.3b). This preferred alignment of triangular seeds reveals that (-201) ledge may represent the preferential nucleation site with the local energetic minimum. With this assumption, we first examine the effect of preferred nucleation sites along the (-201) edges through constructing a cross-sectional atomic model for β-Ga_2O_3 (100). Here, the β-Ga_2O_3 (100) substrate has a monoclinic structure with lattice constants of $\mathbf{a}$ = 3.037 Å, $\mathbf{b}$ = 5.798 Å, and β = 103.8°[109]. Two possible nucleation cases are proposed and their binding energies are calculated: (1) case A, where a Ga atom is notably missing from the vicinal (-201) ledge (Figure 4.3b,c); and (2) case B, whereas Ga atoms remain intact near the (-201) ledge (Appendix B Figure B9). In case A, MoS_2 molecules with 0° and 180° orientations are used as nuclei and are intentionally placed in the vicinity of (-201) edges on β-Ga_2O_3 (100) as schematically represented in Figure 4.3c,d. After relaxation, we find that MoS_2 molecule with 0° orientation predominately docks at the binding sites of (-201) edges.

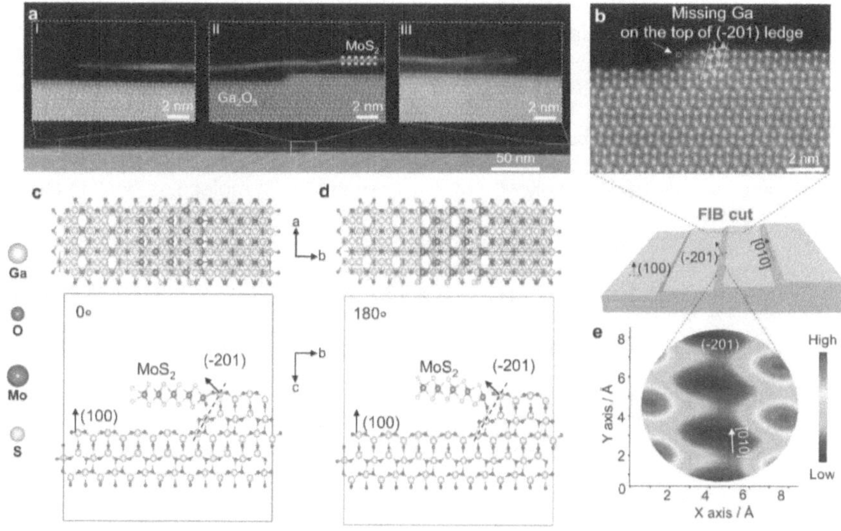

Figure 4.3 Atomically resolved imaging and proposed mechanism for LDE growth of MoS$_2$ nanoribbons on the ledge of β-Ga$_2$O$_3$ (100). (a) Cross-sectional HAADF STEM image of a MoS$_2$ nanoribbon grown on the β-Ga$_2$O$_3$ (100) substrate. The insets provide atomically resolved views of different sections of the MoS$_2$ nanoribbons on: (I) bottom terrace (left); (II) ledge (center); and (III) top terrace (right), respectively. (b) Cross-sectional HAADF HR-STEM image of β-Ga$_2$O$_3$ (100) substrate taken normal to the [010] direction reveals a missing Ga atom from the ledge. Computer-generated atomic models suggest two possible nucleation events on (-201) ledges with orientations toward (c) 0°, and (d) 180°, respectively. The presence of the ledge provides an energetically favorable docking site that breaks the energetic degeneracy by ~2 eV (0°), thus unidirectionally orienting the seeding flakes. (e) Potential energy surface (PES) mapping derived from the DFT calculations sheds light on the diffusion-limited pathway to direct the growth of aligned MoS$_2$ flakes into single-crystal nanoribbons.

Meanwhile, first-principles calculations revealed a drastic difference in the binding energy of ~2 eV relative to that of inversely orientated MoS_2 molecule (180°), thus favoring the mono-oriented growth and therefore the unidirectional alignment. An opposite trend is observed in case B (Appendix B Figure B9), but there exhibits an energy difference of only ~0.535 eV when the MoS_2 molecules dock to the oxygen at the bottom of (-201) ledge. Unlike case A where 0° is the preferred orientation, the preferred orientation in case B is 180° which will lead to mirror grain boundaries in the ribbons (see Appendix B Supplementary Discussion 3 and Figure B10). Certainly, the mono-oriented seeds in our nanoribbons are nucleated following the favorable nucleation case (case A) due to the fact that Ga vacancies are naturally present near the edge of the steps (see Appendix B Figure B11).

The proposed mechanism toward unidirectional nucleation is similar to the recently reported defect-enhanced degeneracy-breaking of TMDs[116] but is quite independent due to the difference in the spatial arrangement of docking sites—randomly distributed and disorganized defect sites vs. spatially ordered and aligned ledge sites. Nevertheless, in this work, they observed the reversal of triangle orientation (i.e., 0° becomes 180°) of MoS_2 flakes across a step edge in the hBN substrate under the assumption of a change in layer polarity of AA' - stacked hBN. On the contrary, the two energetically equivalent but crystallographically inverted ledges, (-201) vs. (001) revealed by DF-STEM and atomic models, across the step edges of β-Ga_2O_3 (100), thus guiding the alignment of MoS_2 nuclei in 0° and 180° orientations respectively (see Appendix B Supplementary Discussion 3 and Figure B11). Once the mono-oriented nucleation approaches completion, the rich sulfur (S) environment not only helps to break the vdW interaction between the aligned MoS_2 seeds and the ledges but also facilities the growth of single-crystalline domains extended beyond both ends of the step edge, ultimately merging together into a continuous nanoribbon.

We note that the growth of individual domains which strongly depends on the diffusion path seems to be confined and directed along the ledges of β-Ga_2O_3 (100). This is very intriguing as the growth of TMDs on highly symmetric substrates by means of a CVD typically results in the omnidirectional diffusion of precursor vapors to the local environment. To verify the origin of this directional diffusion pathway, we performed a potential energy surface (PES) mapping of the (-201) plane of β-Ga_2O_3 (100) via density functional theory (DFT) calculations. As shown in Figure 4.3e, the surface diffusion kinetics along the [010] energetically confine the growth of MoS_2, thus driving the energetically favorable and directionally modulated growth of aligned domains into single-crystalline nanoribbons. These findings collectively point toward an entirely novel strategy to synthesize dense arrays of single-crystalline and globally aligned TMDs monolayer nanoribbons for device applications.

4.2.4 Optical and electrical characterizations

The success of creating extended, single-crystal MoS_2 nanoribbons is manifested in the uninterrupted, homogenous yet narrow distribution of signature PL wavelength across the aligned domains, indicating the lack of atomic misfits between merged domains as shown in Figure 4.4a, and Appendix B Figure B12a.

Meanwhile, hyperspectral PL mapping, which provides a fast, global mapping with a high spatial and spectral resolution, does not reveal any sign of PL quenching typically associated with grain boundaries. Results from conductive (C-) AFM on the MoS_2 nanoribbons directly grown on a semiconducting β-Ga_2O_3 substrate show a similar trend in the representative topography and corresponding current maps are shown in Figure 4.4b. The local point current-voltage (I-Vs, vertical transport) and current mapping were done by applying a positive bias to the β-Ga_2O_3 substrate while the conductive tip (Pt-Ir) was held at the ground (Appendix B Figure B12b,c). The MoS_2 nanoribbons appear highly conducting relative to that of the underlying β-Ga_2O_3 substrate, making them clearly visible in the current map. The average current flowing throughout the MoS_2 nanoribbons in the vertical direction is 18 ($\pm$2) nA. The point I-V curve measured along the MoS_2 nanoribbons exhibits non-ohmic characteristics that appear symmetric. These measurements provide direct experimental evidence of the undisruptive conductive path throughout the entirety of MoS_2 nanoribbons.

Furthermore, we verified the quality of MoS_2 nanoribbons by evaluating the field-effect carrier mobility in a bottom-gate transistor configuration as illustrated in Figure 4.4c and Appendix B Figure B13a-d. To reduce the screening effect from the HfO_2, while eliminating the charge scattering and trap sites, a single-crystal hBN monolayer film is embedded as an interface layer between HfO_2 and MoS_2 nanoribbons[52]. Noted that measurements were performed at room temperature. The electrical properties of our LDE MoS_2 nanoribbons have two important features: the spatial uniformity over a long range similar to those wafer-scale films synthesized by CVD/MOCVD and transport characteristics on par with those seen in exfoliated counterparts. The top panel of Appendix B Figure B13e plots field-effect mobility and on/off ratios measured from five devices, fabricated on the same MoS_2 nanoribbon and separated by up to 20 μm on a single chip. All five FETs exhibit nearly identical behaviors. These include an averaged field-effect mobility of 65 cm^2/V-s and on/off ratios near $\sim 10^8$ independent of channel length and location of MoS_2 nanoribbons, suggesting the spatial homogeneity of the electrical properties of the MoS_2 nanoribbons across various length scales. Specifically, both values are comparable to the performance of the mechanically exfoliated benchmarks as shown in Figure 4.4d[64]. In parallel, the bottom panel of the Appendix B Figure B13e provides the histogram of field-effect mobility and on/off ratios measured from 100 FETs made of different batches of MoS_2 nanoribbons. Evidently, single crystallinity throughout the entirety of MoS_2 nanoribbons is attested by the very narrow distributions of both field-effect mobility and on/off ratios. Occasionally, we find that field-effect mobility of MoS_2 nanoribbons FETs exceeds 100 $cm^2V^{-1}s^{-1}$, with the highest value of 109 $cm^2V^{-1}s^{-1}$ [117]. The mobility is enhanced due to the synergistic effect between the hBN layer and MoS_2 nanoribbons which are both single crystal in nature, ensuring the smooth charge transport along the heterointerface channel (Appendix B Figure B14).

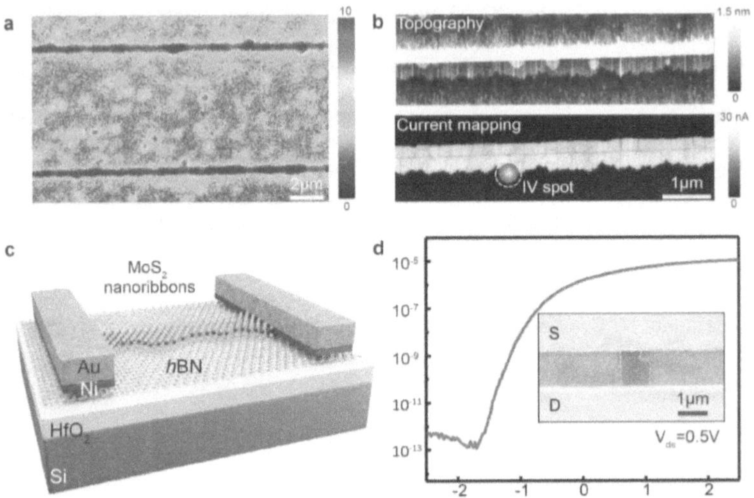

Figure 4.4 Optical, electrical characterizations, and batch fabrication of MoS2 nanoribbon field-effect transistor. (a) Hyper-spectral PL mapping displays a uniform wavelength distribution along the two parallelly aligned MoS$_2$ nanoribbons. (b) Topological (top) and the corresponding current mapping images (bottom) show the spatially uniform current profile of a MoS$_2$ nanoribbon at a constant tip voltage. (c) Schematic illustration shows the device architecture comprised of a monolayer MoS$_2$ nanoribbon stacked on top of a single crystalline hBN interface layer. (d) Transfer characteristic of the MoS$_2$ nanoribbon/hBN field-effect transistor (FET, Inset shows the SEM of a MoS$_2$ nanoribbon sandwiched between source and drain electrodes); the length and width of the device are 1 µm and 0.39 µm, respectively, giving rise to an averaged electron mobility μ = 65 cm^{2}V^{-1}s^{-1} at a drain voltage V$_{ds}$ of 0.5 V.

The location-selective hyperspectral PL included in Appendix B Figure B15a-c revealed that both PL peak positions and FWHM did not vary significantly when the focus of the laser spot was moved across the LDE MoS$_2$ nanoribbon, characteristic of the uniform quality and continuous crystallinity of LDE MoS$_2$ nanoribbons. In parallel, the low-temperature PL (Appendix B Figure B15d,e, excitation: 532 nm, power: 200 µW) shows characteristics unique to exfoliated monolayer MoS_2 benchmarks, including comparable PL intensity, a similar level of defects, neutral exciton and trion emission peaks, further confirming the high quality of LDE MoS_2 nanoribbons. CVD grown-MoS_2 typically exhibits a high density of defects even though these specimens are characterized as high crystallinity. As a consequence, PL induced from defects of CVD-synthesized TMDs emerges and outweighs the intrinsic PL at 4K unless treated chemically or doped electrostatically[118, 119]. The result is the impaired transport property and decreased mobility[120]. This finding of containing a low-level defect density and preserving the high level of single crystallinity simultaneously during the growth stage has not been reported or achieved elsewhere and thus distinguishes the LDE from the other epitaxy approaches.

Because our LDE growth is directed by the combination of ledge sites and surface diffusion-limited pathway, intrinsic to the Ga_2O_3 substrates, its use is not limited to the MoS_2-Ga_2O_3 combination reported here. Instead, it could be generalized for producing various TMDs nanoribbons, including n-(MoS_2), p-(WSe_2), and even lateral n-(MoS_2)-p(WSe_2)-n(MoS_2) junctions with precise single-crystallinity, alignment and monolayer controls over a micro-to-centimeter scale (Appendix B Figure B16). While TMDs nanoribbons with lateral heterostructures have been recently reported by a vapor-liquid-solid growth[27], such a process only allows the growth of heterostructures with either different metals or chalcogen atoms, thus making it challenging for the creation of p-n heterostructures or even n-p-n multi-heterostructures. LDE WSe_2-MoS_2 lateral n-p-n multi-heterojunctions are achieved by growing WSe$_2$ nanoribbons first on β-Ga$_2$O$_3$ (100) substrates, followed by the edge epitaxy of MoS$_2$ nanoribbons on both sides as shown in Appendix B Figure B17a. Hyperspectral PL mapping of relevant PL characteristics, including MoS$_2$ in red, and WSe$_2$ in blue, in tandem with Raman and PL spectra, evidentially proves the successful in-plane growth of n-type MoS$_2$ at both edges of p-type WSe$_2$ (Appendix B Figure B17b-d).

4.3 Conclusion

In summary, we have demonstrated the LDE growth of TMDs nanoribbons. The special structure of the underlying β-Ga$_2$O$_3$ substrate with well-defined steps and relatively large-area ledges allow growing self-aligned, single-crystal large-area, and high-quality monolayer TMDs nanoribbons. This method can be applied to achieve different materials of TMDs and complex heterostructure ready to integrate into nanoelectronic and photonic devices. Our strategy opens the horizon to grow nanoribbons of different materials and can be applied to grow single-crystal TMDs film.

Chapter 5: Aberration Corrected STEM Imaging of 2D Materials: Artifacts and Practical Applications of 3-fold Astigmatism

5.1 Introduction

In contrast to graphene, monolayer two-dimensional (2D) transition metal dichalcogenides (TMDs) exhibit different polymorphs—that is, more than one crystal structure exists. Different crystalline structures of 2D TMDs directly reflect changes in electronic properties, the ability to transport carriers, propensity toward chemical synthesis, and capability to drive catalytic reactions [121, 122]. Unraveling the structure-property-performance relationship requires advanced characterization tools at their native length scales. Imaging characterizations by means of the high resolution scanning transmission electron microscopy (HR-STEM) are gaining unmatched spatial resolution with the development of spherical aberration (Cs) correctors [123, 124]. Particularly, HR-STEM coupled with the annular dark field (ADF) detector, especially in the regime of nearly quadratic dependence of the image contrast on the atomic number (Z-contrast), provides straightforward *in-situ* visualization of the coordination of constituent atoms [125]. Indeed, brighter spots in atomic-scale Z-contrast images of crystalline materials can be directly translated into heavier atoms and vice versa[126]. The discovery of non-monotonic Z dependence, however, raises uncertainty with varying levels of complexity [127]. To prevent the structural misinterpretation of the imaging results requires the integration of the "full package" of model-simulation-experiment pathways (maybe in several iterations), even for Z-contrast HR-STEM [128].

Whereas recent technological advances in the electron-optics design for high-end HR-STEMs [129], have largely negated the adverse effect of dominant aberrations, i.e., Cs of the 3rd order, it remains an open question on how the combination of chromatic aberration (Cc), the 5th orders Cs and non-rotationally symmetric residual aberrations, especially at lower accelerating voltage [130], weighs in and becomes detrimental to the overall (S)TEM resolution [131, 132]. From a theoretic perspective, reaching the highest spatial resolution in STEM requires an electron beam (probe) of an infinitely small size. In practice (due to natural physical limits) [130], this is not achievable and the real electron probe has a definite size, which can be interpreted into the resolution power of a particular STEM instrument.

In modern systems, the probe size can be reduced below 0.1 nm, enabling sub-Å [133] and even sub-50 pm [63] resolution for bulk samples as well as the record-breaking resolution of 0.4 Å [134] for 2D materials. If, at the same time, the probe shape (the beam cross-section intensity distribution) is not spherical but rather triangular in profile, this is described as a 3-fold electron probe astigmatism, often termed as A2 (or a33 in equivalent), where the number stands for the second order of that residual aberration. The value of A2 is measured in nm. STEMs produced by mainstream manufactures are typically equipped with probe Cs-correctors that minimize the A2 below 100 nm, which is considered acceptable.

There are reports with better numbers for A2; however, these numbers are not substantiated by a rigorous measurement that involves statistically significant repetitions and accuracy evaluation. Rather, A2 is obtained from a one-time measurement provided by the corresponding software, which feeds back only the confidence level. This unsystematic approach can lead to unintended consequences that always exist considerable scatter in data, a detriment to evaluating the accuracy. Moreover, for the majority of the STEM systems, the measurement and aberrations tuning algorithm depends on the dedicated alignment of the sample, which is extracted from the microscope before proceeding to the "real" specimen, i.e., the one being actually investigated under STEM. Meanwhile, the value of A2 is known to drift significantly during the course of the characterization and can be influenced easily even through the exchange of specimens.

The effect of electron beam shape on the imaging of 2D layered materials was firstly mentioned by Krivanek et al [135]. In this seminal work, they demonstrated the unmatched ability to commence an atom-by-atom chemical analysis with HR-STEM ADF in 2D layered materials. In particular, the residual aberrations (especially aberrations with the same symmetry as the structure being imaged) can lead to contrast artifacts as a result of the probe 'tails' to the intensity recorded at the nominal probe position [136-138]. This profound effect of residual aberrations on HR-TEM imaging is also discussed in a number of publications [131, 132, 139-141]. However, most of these studies, both for HR-TEM and HR-STEM, predominately focused on graphene, a monolayer of carbon atoms exclusively. This variation in imaging analysis may not cause significant uncertainties when monolayer graphene is being investigated. However, the misinterpretation in imaging analysis can be adverse when 2D monolayer TMDs, which itself contains three layers of atoms and can be polymorphic in nature, such as molybdenum disulfide (MoS_2). Here, the electron beam can be deemed as electromagnetic waves with rotation frequency, e.g., 3-fold astigmatism. The interaction with 2D monolayer TMDs is a consequence of specific phase relationships. When the spatial frequency of the incoming electron beam is "in-phase" with the 2D monolayer TMDs, the tails of the electron beam come across the relatively heavy atoms. The resultant resonant enhancement gives rise to the atomic contrast that scales roughly as the square of the atomic number Z. Meanwhile, other phase relationships are possible between the electron beam and 2D materials that will not lead to this resonant enhancement. When the electron beam is "out-of-phase"—that is, the tails of the electron beam do not constructively interfere/distribute with the 2D monolayer materials. The result is the contrast damping (anti-resonance). It is thus apparent that variation in electron beam orientation can be used to "modulate" the contrast between dissimilar atoms. Together, we term the finding resonant modulation of contrast (RMC) as a recognition that contrast of metal and chalcogenide atoms in HRSTEM images can be modulated by varying the rotation frequency of incoming electron beams.

Here, we report the correlation of atomic simulation packages with HR-STEM ADF imaging, for the first time, to reveal the false T phase of TMDs as a result of the residual 3-fold astigmatism, where its tails can overlap and thus interact with several atomic species simultaneously. This false phase is particularly pronounced for materials made of different atomic species, that arise and then follow a 3-fold

(120°) in-plane rotation symmetry of its origin (*i.e.* A2). Depending on the mutual orientation between A2 and specimen inside the HR-STEM, atomic-scale imaging of the same 2D monolayer TMDs can result in a manually variable contrast, leading to a violation of power-law dependence on atomic number and thus alteration of crystal structure interpretation. Both theoretical simulation and experimental observation suggest that RMC arises even when A2 is around 100 nm and adversely affects the fidelity of structural analysis of 2D monolayer TMDs.

While the advent of an aberration-corrected HR-STEM ADF represents a very powerful yet general analysis tool to determine structural evolution in both real-time and atom-by-atom manners, the combined simulation and experimental characterizations in imaging analyses, allows us to determine the origin and then address the adverse impact of RMC. A viable solution emerges to negate the adverse effect through the employment of an electron beam monochromator, which can achieve an energy spread below 60 meV resolution [61, 62, 142, 143], thus mitigating the RMC below the detection limit and to image 2D TMDs without artifacts regardless of the sample orientation or A2 (up to 150 nm) rotation.

5.2 Methods

5.2.1 Synthesis and characterizations of monolayer transition metal dichalcogenides (TMDs)

MoS_2, WS_2 and WSe_2 films were grown using chemical vapor deposition (CVD) reported previously[9, 144, 145], and were transferred onto the transmission electron microscopy (TEM) grid by a wet transfer method. Optical images and corresponding photoluminescence (PL) mapping were collected using a Witec alpha 300 confocal Raman microscope with a RayShield coupler equipped with a 532-nm wavelength laser.

5.2.2 High-resolution low-loss electron energy loss spectroscopy (EELS)

The measurements were performed at 80 kV with a ThermoFisher USA Titan Themis Z (40-300 kV) TEM equipped with a high brightness electron gun (x-FEG), an electron beam monochromator, and a Gatan Quantum 966 imaging filter (GIF). Spectra were acquired in so-called microprobe STEM mode with about 1 mrad semi convergence angle (4 nm probe size). The monochromator operation was optimized by the method first implemented in ref.[61] and described in detail in ref. [62] to achieve the energy resolution of about 45-50 meV (defined as the full width at half maximum of the zero energy loss peak or ZLP).

5.2.3 X-ray photoelectron spectroscopy (XPS)

The analysis was carried out in a Kratos Axis Supra DLD spectrometer equipped with a monochromatic Al K_α X-ray source (hv = 1486.6 eV) operating at 150 W, a multi-channel plate and delay line detector under a vacuum of ~10-9 mbar. The high-resolution spectrums were collected at fixed analyzer pass energy of 20 eV.

5.2.4 Image Simulations

The image simulations were performed with two software packages employing multi-slice methods: Dr. Probe[146] and QSTEM[147]. These packages allow to build crystal models (including artificial interfaces) using available crystallographic information files (CIFs), and to calculate their TEM and STEM images afterward taking into account the various parameters of TEM (high tension, aberrations, instabilities, etc.) as well as the orientation of the specimen. Atomic models of synthetic grain boundaries formed by domains with a misorientation of 180° (MoS_2) and 90° (WS_2) degrees were constructed using QSTEM Model Builder and Vesta software[148]. We also use available Digital Micrograph (Gatan, USA) scripts to simulate noise for the best match to experimental HR-STEM imaging. Simulation parameters were chosen as listed in Appendix C Table C1.

5.2.5 HR-STEM ADF

HR-STEM ADF imaging was performed with a ThermoFisher Titan Themis Z (40-300 kV) TEM equipped with a double Cs (spherical aberration) corrector, a high brightness electron gun (x-FEG), and an electron beam monochromator. To reduce the electron beam sample damage, we chose to operate the microscope at 80 kV and tune it to minimize 3rd order Cs below the detection limit (few μm). However, some residual 3-fold astigmatism of about 100 nm (as measured by the Thermofisher Cs-Probe corrector software with a standard alignment sample) is intentionally left uncorrected. Images were acquired in diffraction mode with the camera length 115 mm corresponding to collection angles of 41 (inner) and 200 (outer) mrad of a Fischione Annular Dark Filed detector. Probe semi-convergence angle was tuned for 30 mrad with the beam current ranging from 50 pA for the regular STEM to 3 pA for the monochromated beam STEM. For the monochromator operation, the method first implemented in ref.[61] and described in detail in ref. [62] was used. As a result, the monochromatic STEM was performed with the energy spread in the beam of about 60 meV (defined as the full width at half maximum of the zero-energy loss peak). Gauss high-pass (to reduce contamination effects) and low-pass (to reduce scanning noise) filtering was used to enhance the contrast of the image [63] (Appendix C Figure C2-C6 and C12-C16). Measurements and calibration of optical elements of Cs probe corrector to adjust the A2 astigmatism were performed by standard ThermoFisher software provided with a Cs probe corrector and by the use of a standard alignment sample (Cross Grating Replica 3 mm, AGS106). A Fischione Dual-Axis Tomography Holder (model 2040) is used as this holder allows the specimen to be fully rotated through 360° in the plane orthogonal to the electron beam.

5.3 Results and Discussion

5.3.1 TMDs mirror grain boundary: A model system

To demonstrate the RMC, 2D monolayer MoS_2 was used as a representative model system which is known for exhibiting a variety of polymorphs because an individual MoS_2 monolayer, which itself comprises three layers of atoms (S-Mo-S), can be in either one of the two phases, namely 2H and 1T polymorphs (Appendix C Figure C1). Although both domains in the mirror grain boundaries are made of intrinsic 2H phases, S orientation with respect to Mo on both domains is not spatially equivalent as shown in Figure 5.1a. On the left-hand side—namely, LHS, of the model, each pair of S atoms adjacent to Mo in the vertical direction is located above Mo. For clarity, we denote such a configuration as 2H (S↑) and mark it with the green arrow up. On the right-hand side, termed RHS, the orientation is reversed where two S atoms are now positioned below Mo. Therefore, this is assigned as the 2H (S↓) configuration, highlighted with the green arrow down. The 2H (S↑) and 2H (S↓) configurations are differentiated only by a 60° or 180° in-plane rotation. This inequivalence in the in-plane orientation provides a well-defined model system that gives rise to a stark atomic contrast in ADF imaging between opposite domains. Such a model of 2H (S↑) - 2H (S↓) mirror grain boundary corresponds to the typical growth orientation mismatch in monolayer MoS_2 films grown by CVD [144].

Since A2 is the aberration with a strong rotation anisotropy, it is assumed that imaging artifacts will hinge strongly on the orientation of the specimen relative to A2. For the clarity of notation, the orientation of A2 (with respect to the model structure) with one of the corners pointing up is defined as (A2↑). The orientations of (A2↑) and (A2↓) are the result of 60° rotation while rotating by 30° affords (A2←) and (A2→). The relative magnitude of A2 increases with increasing lengths of red arrows. Depending on the mutual orientation between MoS_2 and A2, RMC could arise and influence the imaging results, and hence the structural interpretation as shown in Figure 5.1b and c.

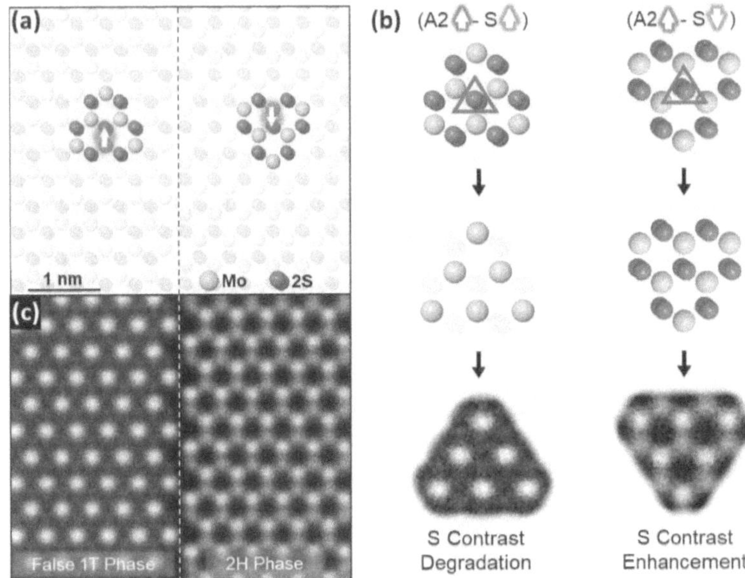

Figure 5.1 Schematic illustration of the emergence of RMC as a result of the mutual orientations between A2 (red arrow) and MoS₂ (green arrow). (a) Computer-generated atomic models of monolayer MoS$_2$ with a mirror grain boundary. (b) Schematic representations of the dissimilar atomic configurations where S contrast degrades on LHS and enhancement in S contrast is observed on RHS. (c) Simulation of HR-STEM ADF imaging for the atomic models in (a) with A2 applied.

5.3.2 Simulation

As a first step in investigating the role of RMC in imaging of 2D monolayer MoS$_2$, Dr. Probe [146] and QSTEM [147] software packages were employed to simulated HR-STEM ADF of the mirror grain boundary, across which spatial orientations alter, e.g., 2H (S↑) - 2H (S↓), Figure 5.2a. Conditions implemented are based on the model in Figure 5.2b, for the case of Cs =1 μm and all the residual aberrations, including A2, set to zero. On both sides of the mirror grain boundary, S and Mo atoms are clearly resolved: 3 Mo atoms coordinated with 3 pairs of S atoms form a 6-fold ring, characteristic of the 2H phase. Next, we gradually (50 nm increment) increase A2 in our simulations. The effect of A2 on the electron beam shape is shown in the insets of Figure 5.2a, d-l. For A2=0 nm, nearly all electrons are concentrated at the center of the beam, giving rise to a round-shaped probe. With increased magnitude of A2, significant numbers of electrons are redistributed toward the tails, resulting in a triangular-shaped and much bigger probe. Our simulations reveal that increasing a 3-fold astigmatism in the orientation (A2↓) on the RHS (S↓) is manifested in the dominant emergence of Mo atoms, while the contrast of S atoms drastically decreases (Figure 5.2d-f), compared with the LHS (S↑). Already, at A2=100 nm the contrast of S on the (S↓) side nearly vanishes. Meanwhile, in the case of RHS (S↓) shown in Figure 5.2e and f, the "visual transition" from the 2H (S↑) - 2H (S↓) mirror grain boundary to the 2H - false 1T (false 1T refers to 1T-like phase caused by the RMC) phase boundary takes hold at A2=150 nm, closely resembling the intrinsic 1T or "S contrast degradation" often reported for the metallic 1T MoS_2.

As indicated in the model in Appendix C Figure C1b, each S site in the projection of the 2D crystal lattice comprises only one atom (as opposite to 2H phase with two S atoms in the same projection) whose exceedingly low signal-to-noise ratio makes it indiscernible from HR-STEM ADF imaging, leaving behind discernable patterns of Mo atoms. While such a glaring contrast in HR-STEM ADF imaging has long been used to unambiguously distinguish these chemically homogeneous but crystallographically heterogeneous phases, it becomes a formidable challenge to determine a given domain whether it is an artificially false 1T or an intrinsically metallic 1T phase. As schematically represented in Figure 5.2c, when the electron probe (red triangle) is arranged in the (A2↓- S↑) configuration, the 3-fold tails of the probe are found to overlap with the neighboring Mo atoms and hence lead to stronger scattering of the electron beam. The scatter in the electron beam is a consequence of the heavier Mo atoms— that is, higher contrast in HR-STEM ADF. Even a small portion of the beam (tails) scattered at this position by Mo atoms will inevitably add up to the S signal, concurrently enhancing the effective contrast of (S↑) atoms in LHS (Figure 5.2d-f) and thus the emergence of the 2H phase.

On the other hand, for the (A2↓- S↓, Figure 5.2c) configuration, when the electron probe is placed on S, its tails will not interfere with Mo atoms and thus no enhancement in S contrast is observed. Similar to the intrinsic 1T phase, this lack of enhancement of S contrast also renders S "invisible", leading to the inception of a false 1T type of contrast. Likewise, "visual transition" from false 1T→2H is demonstrated in Figure 5.2g-i when reversing the direction of A2 (A2↓ to A2↑) while keeping the S configuration constant. Meanwhile, it is also noted that there is an intermediate state (A2←) or (A2→), for which contrast of S atoms degrades (due to the effectively bigger probe) equally on both sides of the domains (Figure 5.2j-l). Drawing on these simulation results, a picture emerges indicating that RMC of lighter elements where electron probe tails intersect neighboring heavier atoms, comes into effect when 3-fold astigmatism is in the range of 100-150 nm coupled with the direction of A2 relative to the crystal structure orientation.

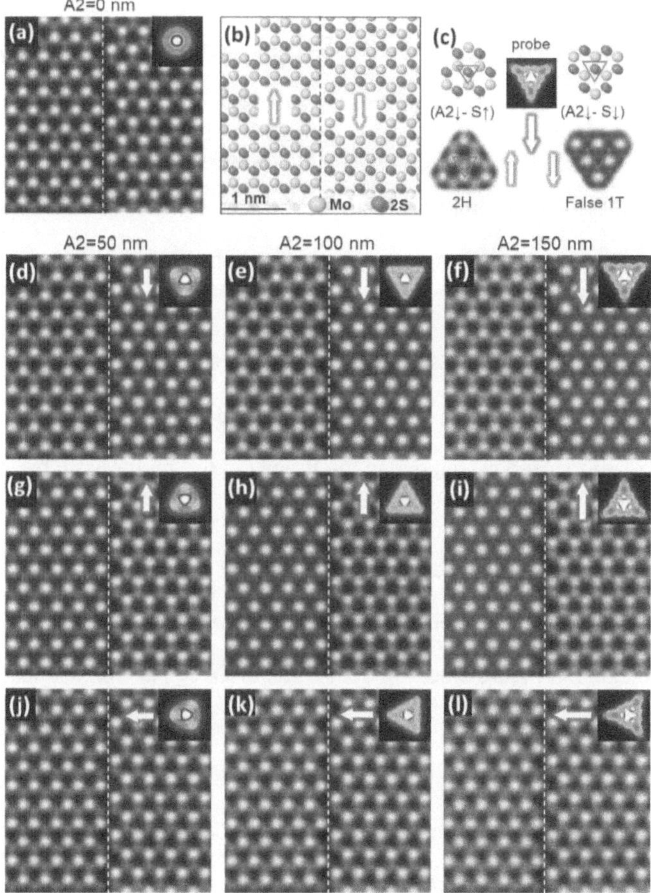

Figure 5.2 QSTEM Simulation of HR-STEM ADF imaging. Domains of 2D monolayer MoS2 separated with antiphase (60° or 180° in-plane rotation) growth mismatch for Cs corrected electron probe with various A2. (a) Image taken from QSTEM simulation based on the atomic model in (B) with A2=0 nm. (d-l) Same as (a) but with different values and orientations of A2 introduced. The insets in (a) and (d-l) show the relative magnitude and orientations of A2 (red arrow), and the attendant evolution of the electron probe shape. (c) Schematics of A2 orientation effect on S contrast.

5.3.3 Experiment

To confirm the predictions from QSTEM simulations, 2D monolayer MoS_2 with a 2H ($S\uparrow$) - 2H ($S\downarrow$) mirror grain boundary with a 60° growth mismatch (see supplementary discussion) was synthesized via chemical vapor deposition (CVD). HR-STEM ADF imaging was performed to resolve the atomic structures specifically on regions separated by a sharp grain boundary with the following conditions: (i) manual in-plane rotation of the 2H MoS_2 specimen inside the TEM holder for otherwise fixed microscope conditions, followed by (ii) imposing and rotating a 3-fold astigmatism with respect to a fixed sample orientation. To this end, we first focused on regions in the vicinity of the mirror grain boundary where the two MoS_2 domains intersect as schematically illustrated in Figure 5.3.

First, the TEM holder was rotated clockwise *in-situ* with a 10° increment until the RMC is at its peak on one side, and hence, false 1T phase is observed. As indicated in Figure 5.3b, the domain on LHS ($S\uparrow$) is characterized by the trigonal prismatic coordination, e.g., 2H phase, whereas the significantly reduced contrast of S atoms on RHS ($S\downarrow$) appears as the false 1T when A2 is applied. The presence of 2H—false 1T polymorphs matches well with the simulation results shown in Figure 5.3e and f. Indeed, after manually rotating the double truncated MoS_2 specimen by 60° counter-clockwise (Figure 5.3c), we again observed a reversion to the false 1T—2H polymorphs as is evident in Figure 5.3d. The reversion of contrast that alternates in every 60° increment confirms the 3-fold symmetry of the RMC. Importantly, these observations are in good agreement with theoretical modeling and therefore allow us to determine the orientation of A2 during the HR-

STEM ADF imaging.

Another important factor that weighs in the visual transition between 2H and false 1T is the direction of A2. Here, we rotated the direction of A2 through the assistance of a Cs-probe corrector software (see details in SI) until the maximum RMC was achieved and hence the emergence of visual transition from 2H→false 1T. As shown in Figure 5.3e, 2H domain (S↑) above the abrupt phase boundary is unambiguously juxtaposed with the false 1T phase in the bottom domain (S↓), corroborating the existence of RMC. In the next step, A2 is rotated by 180° (60°). Such rotation of A2 results in the inversion of the contrast across the phase boundary and thus reverses the phases on both domains as demonstrated in Figure 5.3f.

Meanwhile, point defects in CVD-grown MoS_2, including substitutional, interstitial atoms, and vacancies, have been reported by several groups and are known to induce pronounced structural distortions as highlighted with yellow arrows in both Figure 5.3e and f. If phase transformation truly took place, top and bottom planes of S atoms are forced to shift away. This may, in turn, propel structural defects deviated away from their origins slightly although a more vigorous investigation is required. We surmise that point defects should persist if phase reversion is truly artificial in nature as opposed to rearrangement of lattice or sliding of atomic planes. Here, the point defect is, however, found to remain still in its original position and differs only by the atomic contrast after coupling with the probe tails interaction and the presence of RMC. The persistence of point defects confirms that no major structural rearrangement or atomic displacement occurs during the

A2 rotation. To further verify the influence of the magnitude of A2 on phase transition, the magnitude of A2 in Figure 5.3g is reduced to 50 nm (vs. 100 nm in Figure 5.3e), which is below the detection limit. The influence of A2 fades quickly and the 6-fold atomic rings that constitute the 2H phase again come into sight on both domains despite the relatively low S contrast in accordance with the simulation results featured in Figure 5.2a.

5.3.4 Mitigating the RMC

Yet, the finding here also signals the need for searching alternative ways to improve the atomic contrast and separation (i.e., resolution) without the interference of RMC. We demonstrate that the implementation of a monochromated beam can effectively address the RMC (Figure 5.3h) while improving the resolution [61, 62, 142, 143]. Making use of the monochromator allows us to achieve the energy spread ΔE in the beam of ~60 meV (determined from the full width at half maximum of the zero-energy loss peak) compared to 1 eV in the original beam. With a typical value of chromatic aberration for TEM of Cc=1.7 mm, the "energy length" parameter Cc ΔE is then maintained well under 0.1 eV mm. Consequently, the effect of chromatic aberrations of the electron-optic system is drastically reduced, allowing further demagnification of the probe while significantly improving the spatial resolution by nearly 200% [130] as compared between Figure 5.3g and h, respectively.

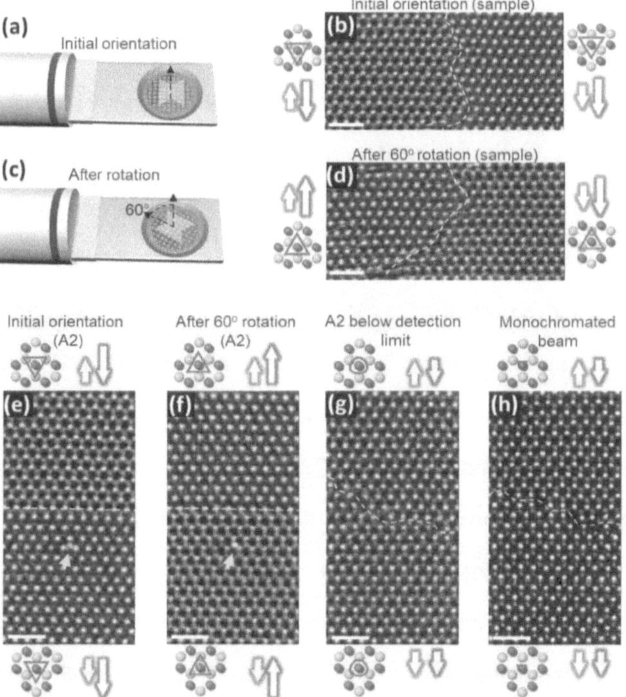

Figure 5.3 Experimental observations of RMC in HR-STEM ADF imaging. (a-d) Sample rotation experiment: HRSTEM-ADF imaging taken at regions in close proximity to the interface between MoS_2 mirror grains with antiphase (60°) growth mismatch. Schematic representation of the relative position of sample supported on TEM holder before and after rotation is illustrated in (a) and (c). Corresponding HR-STEM ADF images before and after an in-plane rotation by 60° are shown in (b) and (d). The experimental observation agrees very well with the QSTEM simulation results as is evident by the atomically resolved phase transition after A2 rotation. (e-h) A2 rotation experiment: Cs-corrected HRSTEM-ADF imaging of the interface between a monolayer MoS_2 grains with antiphase (60°) growth mismatch. (e) and (f) Images at the same spot of the interface before and after A2=100 nm rotation by 60°, correspondingly. Atomic models (Mo, color in yellow and S, color in green) and arrows (green for sulfur, red for A2) suggest A2 orientation based on the simulation in Figure 5.2. (g) Image of the interface for A2 reduced below the detection limit and (h) additionally, a monochromated beam (60 meV energy spread) is employed. Scale bar is 1 nm.

5.3.5 Phase characterization

More insights into the structural continuity and chemical composition of CVD-grown MoS_2 are gained through combined spectroscopic characterizations, including photoluminescence (PL) mapping, electron energy loss spectroscopy (EELS) bandgap measurements, and X-ray photoelectron spectroscopy (XPS). To ensure the coherence and validity of data interpretation, the same MoS_2 specimen characterized with the 2H (S↑) - 2H (S↓) mirror grain boundary was directly used without post-transferring from the TEM grid. Appendix C Figure C8a features an optical image of a butterfly-shaped MoS_2. Corresponding PL intensity mapping shown in Appendix C Figure C8b closely resembles the optical image with a uniform yet narrow wavelength distribution intensity in both domains, indicating that both domains are indeed semiconducting (2H phase) in nature. In parallel, EELS bandgap measurement on both domains further suggests that conduction band onset arises around 1.55 eV (Appendix C Figure C8c) and displays a linear increase of the intensity, which is characteristic of 2D monolayer MoS_2 [62]. Meanwhile, XPS spectra of Mo 3d core level electrons show two characteristic peaks assigned at 229.7 and 232.8 eV from Mo $3d_{5/2}$ and Mo $3d_{5/2}$ in Mo (IV), originating from 2H MoS_2 (Appendix C Figure C8d) [149, 150]. Overall, these characterizations collectively confirm that both domains of the butterfly-shaped MoS_2 flake are indeed intrinsic 2H phases and the appearance of false 1T phase is, in fact, the result of a violation of the power-law dependence of contrast on coordination modes between the transition metal and chalcogenide atoms.

In addition to the mismatch in domain orientations, CVD-grown monolayer MoS_2 is also known to undergo the localized phase transformation induced by electron beams, and often resolved as a 2H→1T phase transition [59, 151]. To discriminate the innate T phase from the false T phase, we again put the MoS_2 specimen with a 2H (S↑)-2H (S↓) mirror grain boundary into a test. When applied ~100 nm of intentionally induced 3-fold astigmatism, LHS (A2↓- S↑) highlighted in red shows the 2H phase while the RHS (A2↓- S↓) highlighted in yellow shows the false 1T phase as suggested in Figure 5.4a. Aside from the mismatch of global domain orientations as shown in Figure 5.2, here we also demonstrate that local transition (embedded patch highlighted with red color in RHS) is induced by the electron beam during the image focus optimization and 2-fold astigmatism adjustment. In this patch, the S atoms are clearly resolved in S↑ position (as a result of Mo plane or both S planes gliding [59], Figure 5.4c), *i.e.,* the patch emerges as a 2H phase and is surrounded by the matrix of a false 1T phase. Meanwhile, as A2 is deliberately rotated by 60°, the image contrast reversion again occurs universally across both domains as well as taking place locally at the embedded patch (Figure 5.4b). Note that the patch is not only increased in dimensions by virtue of an extra beam exposure but also reverted back to the false 1T phase (now surrounded by a 2H matrix). Observing such a contrast reversion with the A2 rotation which carries out both universally and locally, indicates that electron beam-induced 2H (S↑)→2H (S↓) transition (attributed to Mo plane or both S planes gliding) mixed with RMC are the reasons of the false 1T phase observed commonly in HR-STEM ADF images.

Additionally, we demonstrate that the presence of RMC can be repurposed to facilitate the identification of S vacancies. The diminished contrast of S atoms during the RMC-driven 2H→false 1T transitions (RHS with A2↓-S↓ configuration, the false 1T) makes the S vacancies relatively visible when compared to the original 2H phase (Appendix C Figure C10). This characteristic feature combined with the intensity profile may be leveraged as a straightforward tool for revealing structural defects with vivid clarity as shown in Figure 5.4 a and b.

The combinatory studies that synergistically integrate complementary strengths from both QSTEM simulation and HR-STEM ADF imaging characterizations indicate the manifestation of RMC when the difference in atomic numbers between coordination atoms is pronounced at the presence of A2. To prove the profound and generic implication of RMC, an array of representative 2D monolayer TMDs, including WS_2, WSe_2, and graphene, have been put into test and all of them displayed RMC-dependent atomic phase-contrast albeit at various degrees (see Appendix C Figure C11-C18 for other systems). Moreover, false atomic contrast emanated from RMC deviates from the power-law dependence on the specimen atomic composition. This deviation from the classic power law coupled with the magnitude of A2 suggests that lighter atoms, such as Se, can appear almost as bright as nearby heavier elements (W), leading to a complete misinterpretation of HR-STEM ADF imaging results (Appendix C Figure C19).

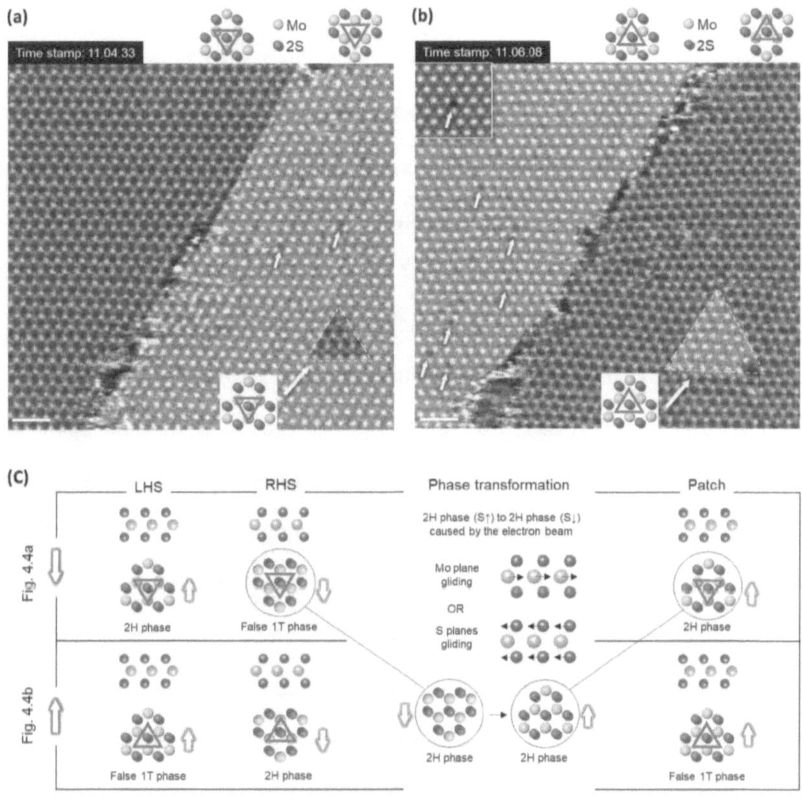

Figure 5.4 Electron beam induced local transition of embedded patch inside the MoS₂ matrix. False-colored HR-STEM ADF images at a mirror grain boundary. (a) before, and (b) after A2=100 nm and rotating by 60°, respectively. Color coding is used to distinguish regions with 2H (red) and false 1T (yellow) types of contrast, caused by (A2↓- S↑) and (A2↑- S↑) imaging configuration, respectively. Red- and yellow triangles in (a) and (b) show the growing 2H (S↑) transition region in the 2H (S↓) matrix. Note that white arrows denote double S vacancies sites. The inset in (b) shows the simulation of false 1T contrast of MoS₂ film with a double S-vacancy in the center. Scale bar is 1 nm. (c) Schematic representations reveal the root cause of contrast formation in both (a) and (b).

5.4 Conclusion

The present finding illuminates the fundamental roles of the atomic contrast, spatial inequivalent S orientation, and residual 3-fold astigmatism on the determination of the atomic structure of 2D monolayer TMDs. Corroborating QSTEM simulations with the experimental observation of HR-STEM ADF imaging of 2D monolayer TMDs allows us to reveal the artificial nature of atomic-scale phase transitions with a relatively small amount of residual low order aberration, (A2=100-150 nm). The presence of RMC can result in artificial structures extrinsic to the specimen in HR-STEM ADF imaging both selectively and generally. Importantly, these results should act as a reminder to researchers engaging in emerging 2D materials research: in addition to utilizing HR-STEM ADF imaging to probe atomic arrangements and predict associated material properties, it is perhaps even more important to rule out the possible interference of the extended tails of the 3-fold astigmatic electron beam that can interact with neighboring atoms to induce unwanted RMC. We suggest employing more critical approaches for interpretation of HR-STEM ADF data, especially in 2D materials. The employment of an electron beam monochromator allows us to mitigate the RMC below the detection limit and to image 2D monolayer TMDs and beyond without artifacts regardless of the sample orientation or A2 (up to 150 nm) rotation.

Chapter 6: Conclusions and Future Direction:

6.1 Conclusion:

To conclude this book, controlling the growth of 2D TMDs is a necessity to ensure the full potential of these materials to reach the industrial level for mass production. We have comprehensively studied the factors that affect the growth and orientation of the domains in the gas-phase and solid-phase (on a substrate) which are the building blocks of the large-scale production of non-silicon materials.

In Chapter 2, the effect of the precursors' concentration ratios on the growth is investigated which gives a better understanding of controlling the composition of the seeds in the gas phase. As a result, aligned growth is achieved over the whole substrate that scans the symmetry of the substrate.

In Chapter 3, an asymmetrical substrate with naturally-accruing ledges is used to break the 3-fold symmetry that results in mirror grain boundaries, and as a result, single orientation domains were achieved. Moreover, the unidirectional domains are stitched and formed nanoribbons following the direction of the ledge.

In Chapter 4, the relation between the 3-fold symmetry of the 2D TMDs and residual 3-fold astigmatism during HR-STEM ADF imaging is studied along with the artifacts which can lead to misinterpretation of the imaging results.

www.ingramcontent.com/pod-product-compliance
Lightning Source LLC
LaVergne TN
LVHW041713190726
843493LV00007B/2062